Metropolis, Money and Markets

This book explores the impact of finance on urban spaces as well as cities' role in the social constitution and dissemination of financial logics and techniques. It brings together literature from different disciplinary areas to increase our understanding of financialization.

It observes how non-financial members of society, such as public bureaucrats, urban planners, the media and so on, are actively involved in the financialization of urban areas. With an explicit focus on Brazil, a developing country in the Global South, the book demonstrates how the country has been grappling with complex and contradictory processes of neoliberalization, decentralization, re-democratization and institutional-legal strengthening of frameworks for urban and regional planning, stressing the relations between urban space and finance capital.

With a distinct view of filling a gap in the current literature on urban financialization, the book aims to focus on less developed areas in this field and link them with the literature on social studies of finance. This makes the text relevant for academics and scholars of urban studies and planning theory, geography, development studies and political economy, as well as scholars in the US and Europe interested in understanding Brazilian patterns of financialization.

Jeroen Klink is a Dutch economist who works as a professor of economics and urban planning at the Universidade Federal do ABC (São Paulo, Brazil). He has published several books and papers on metropolitan governance, economic restructuring in cities and the role of finance.

Global Urban Studies
Series Editor: Laura Reese
Michigan State University, USA

This series is published in conjunction with the School of Planning, Design, and Construction at Michigan State University, USA.

Providing cutting edge interdisciplinary research on spatial, political, cultural and economic processes and issues in urban areas across the US and the world, books in this series examine the global processes that impact and unite urban areas. The organizing theme of the book series is the reality that behavior within and between cities and urban regions must be understood in a larger domestic and international context. An explicitly comparative approach to understanding urban issues and problems allows scholars and students to consider and analyse new ways in which urban areas across different societies and within the same society interact with each other and address a common set of challenges or issues. Books in the series cover topics which are common to urban areas globally, yet illustrate the similarities and differences in conditions, approaches, and solutions across the world, such as environment/brownfields, sustainability, health, economic development, culture, governance and national security. In short, the Global Urban Studies book series takes an interdisciplinary approach to emergent urban issues using a global or comparative perspective.

Twin Cities
Urban Communities, Borders and Relationships over Time
Edited by John Garrard and Ekaterina Mikhailova

Disassembled Cities
Social and Spatial Strategies to Reassemble Communities
Edited by Elizabeth L. Sweet

Metropolis, Money and Markets
Brazilian Urban Financialization in Times of Re-emerging Global Finance
Jeroen Klink

For more information about this series, please visit: www.routledge.com/Global-Urban-Studies/book-series/ASHSER-1385

Metropolis, Money and Markets

Brazilian Urban Financialization
in Times of Re-emerging Global Finance

Jeroen Klink

Routledge
Taylor & Francis Group

LONDON AND NEW YORK

First published 2020
by Routledge
2 Park Square, Milton Park, Abingdon, Oxon OX14 4RN

and by Routledge
605 Third Avenue, New York, NY 10017

Routledge is an imprint of the Taylor & Francis Group, an informa business

First issued in paperback 2021

British Library Cataloguing-in-Publication Data
A catalogue record for this book is available from the British Library

Library of Congress Cataloging-in-Publication Data
A catalog record for this book has been requested

ISBN 13: 978-0-367-18078-2 (hbk)
ISBN 13: 978-0-429-05943-8 (ebk)
ISBN 13: 978-1-03-223836-4 (pbk)

DOI: 10.4324/9780429059438

Typeset in Times New Roman
by Apex CoVantage, LLC

Contents

Acknowledgements   vi

Introduction   1

1   A conceptual primer on (urban) financialization   16

2   State spaces, financialization of cities and the urbanization
    of finance in the Global South   51

3   State spaces, urban rents and finance in the Brazilian
    development trajectory   82

4   Planning, projects and profitability: The role of financial
    and institutional devices in the urbanization of finance   107

5   On contested water governance and the making of urban
    financialization: Exploring the case of metropolitan
    São Paulo, Brazil   131

6   Building within the limits? A closer look at the "My House,
    My Life" Program in the outskirts of metropolitan São Paulo   148

7   On reframing the capital market–austerity nexus
    and the emerging governance of urban securitization   186

    Conclusion   208

    Index   216

Acknowledgements

Sections 4.2 and 4.3 are reprinted from the *Land Use Policy, Volume, 69*, journal, Jeroen Klink and Lais Eleonora Maróstica Stroher, "The Making of Urban Financialization? An exploration of Brazilian urban partnership operations with building certificates", pages 521–526 (2017), published by Elsevier. Sections 5.2, 5.3 and 5.4 first appeared online in the journal, *Urban Studies*, Jeroen Klink, Vanessa Lucena Empinotti and Marcelo Aversa, "On contested water governance and the making of urban financialization: Exploring the case of metropolitan São Paulo, Brazil, pages 6–20 (2019), published by Sage. I would like to thank the editors and publishers of these journals for permission to republish part of these articles here. Figure 5.1 (Evolution of nominal share price of SABESP (2002–2018)) is taken with permission from Lucas Daniel Ferreira. Finally, the research on which part of this book is based received support from the agency *Conselho Nacional de Desenvolvimento Científico e Tecnológico* (CNPQ) (i.e. National Council for Scientific and Technological Development), through Grant No. 307069/2017–6 from the Brazilian Ministry for Science, Technology, Innovation and Communication.

Introduction

Particularly since the 2007/08 US subprime mortgage meltdown, which triggered what can probably be considered as one of the worst crises in the world economy since the 1929 crash, academic research on financialization has proliferated rapidly. Aalbers (2015: 214) has defined the concept as "the increasing dominance of financial actors, markets, practices, measurements and narratives, at various scales, resulting in a structural transformation of economies, firms (including financial institutions), states and households".

Although the variety in methodological approaches and disciplinary perspectives makes it difficult to generalize, it is nevertheless possible to detect specific strands in the literature on financialization (Zwan, 2014), such as Marxian, regulationist or Post-Keynesian political economy (e.g. see Hilferding, 1910; Chesnais, 1997; and Minsky, 1996, respectively); critical management studies and shareholder governance (Froud et al., 2006); social studies of finance, including work on the performativity of economic-financial models (Mackenzie, 2006; Berndt and Boeckler, 2009); and financial geography (French et al., 2011).

Political economy has emphasized structural transformations in the generation of a finance-driven mode of accumulation and regulation that "exploits all of us without producing" (Lapavitsas, 2013).

Work on critical management studies and shareholder governance has used corporate finance and principal-agent theory (Fama, 1980) in order to flesh out the contradictory reshaping of organizations and governance under the influence of shareholder value, structured around the premise of maximizing capital gains and dividends (Froud et al., 2002; Erturk et al., 2007).

Social studies of finance have prioritized the investigation of the mechanisms through which finance has penetrated into the reproduction of daily life, for example, through credit cards, small-scale saving schemes and micro-finance, consumer credit, mortgages and private health and retirement plans. This literature describes how the hollowing out of welfare regimes has shifted responsibilities and insurances from the state to the investing subject. The focus becomes on how, particularly in relation to non-elite actors, "risk itself becomes the motivating force to enter financial markets for protection against possible unemployment, poor health or retirement" (Zwan, 2014: 112). For Martin (2002), life itself has become an asset that requires to be managed. Related research has emerged on how individual subjectivities and the circulation of norms, expectations and conventions in social networks have shaped collective calculative practices and the pricing and valuation in housing markets (Munro and Smith, 2008). Along similar lines, work on performativity has argued that economic and financial theories, models and techniques not only describe the reality of the stock exchange but potentially contribute to constitute and format these markets (Callon, 1998; Mackenzie, 2006). Similar investigations have indicated a gradual financialization of valuation and pricing practices as well. Chiapello (2015), for instance, discusses how a key metric used in financial economics for the (e)valuation of projects and programs (i.e. the net present value of expected flows of income associated with a particular asset) has changed traditional "historical-cost" accounting practices for assets into a valuation approach whereby asset prices instead reflect their "projected" future earning capacity (see Chapters 1 and 2).

Finally, financial geography has stressed the variegated and intrinsically uneven spatial character of financialization whereby "money flows like mercury", "pooling" in particular places and spaces (Hall, 2010: 240). Aalbers (2019: 2) argues that, while its traditional concern, "as a sub-sub-discipline embedded in economic geography", has been "the research of financial centres and districts", financial geography has increasingly stretched "its frontier to the intersection of economic and urban geography".

This book aims to provide at least three interrelated contributions to this growing literature on finance increasingly dominating the economy, society and daily life.

First, and working along the lines of some of the sceptics such as Christophers (2015), much research has lost political relevance through its rather "reactive focus" by assuming financialization as a starting point and investigating its consequences in terms of "what finance does" in the daily life of others, mostly non-financial actors, through the extraction of value without producing. However, in this book we broaden the research focus by trying to increase our understanding of "what finance is" and how it is socially and politically constituted. To be specific, such a perspective doesn't deny the existence of broader patterns marked by an increasing penetration of finance into the economy, politics and society but allows for a closer articulation between political economy – with its priority on the generation and extraction of value, while, by and large, taking markets for granted – and more open-ended social studies of finance-inspired work (Christophers, 2014). The latter is focussed on fleshing out "the black box of finance" itself, including the metrics, collective agencies and circulation of norms and conventions in professional communities. This involves analyzing the entanglements between financial and non-financial actors alike in the design of pricing and valuation practices that accompany "the making of" markets and financialization itself. Moreover, such an approach has the potential to increase the political relevance of research on financialization by providing fresh entrance points and ammunition for the design of counterhegemonic projects and strategies aimed at what Polanyi (2001) called the re-embedding of deregulated financial markets back into society through a "robust palimpsest of social, cultural, political, and regulatory constraints and controls" (Kirkpatrick, 2016: 56; Fields, 2014).

***

A second contribution of this book is the development of an analytical story line and associated empirical work on financialization as a specifically urban-metropolitan phenomenon. This implies going beyond financialization *in* cities towards studying financialization *of* cities and *through* cities in times of re-emerging global finance. To be clear, we don't deny the literature has recognized that financialization has effected and is also influenced by cities in the post-Bretton Woods setting (Theurillat and Vera-Büchel, 2016; O'Neil, 2018). It has engaged with at least four interconnected themes (i.e. financialization of urban land; entrepreneurialism, urban governance and large urban development projects; infrastructure financialization; and the city-capital market nexus).

Although Harvey (1982), based on Marx, already in the 1980s analyzed the tendency of land to be transformed into a financial asset, more recent literature has now started to flesh out the specific geographic and historical circumstances under which this occurs, including an analysis on the limits of the process itself and specific case studies on countries such as Singapore (Haila, 1988) and the Asian developmental states (Haila, 2016; Aveline-Dubach, 2016). Other authors have examined the role of the state in the financialization of public land (e.g. Christophers, 2017 for the UK case) or how land banks have been used within corporate strategies (e.g. see Sanfelici, 2013 on how Brazilian developers used land banks in order to finance the speculative expansion of their market shares).

A second theme (i.e. the articulations between urban neoliberalization and large urban redevelopment projects) is also not new in the literature (Brenner, 2000; Swyngedouw et al., 2002). Nevertheless, recent work has innovated with an analysis on how the entanglements between financial and non-financial actors alike in redevelopment projects tend to transform city space into a "portfolio of tradable income yielding assets" (Guironnet and Halbert, 2015). This has included growing empirical evidence on the role of finance in the design and implementation of projects (Kaika and Rugierro, 2016; Charnock et al., 2014) as well as in the financialization of entrepreneurialism itself through the involvement of the (local) state in mobilizing finance for the delivery of land, infrastructure and real estate (Savini and Aalbers, 2016; Charnock et al., 2014; Klink and Stroher, 2017).

In relation to the third theme, that is, infrastructure finance, the entrance point of the literature has been Graham and Marvin's (2001) path-breaking work on state rescaling, neoliberal infrastructure transitions and splintering urbanism. A common thread in the discussions has been the "relational role" of finance in generating innovative products aimed at the repackaging of illiquid, high-risk and lumpy infrastructure networks into desirable investment assets, and receiving a premium for exactly doing that (Pryke and Allen, 2017; Torrance, 2008). Nevertheless, the unbundling, disassembling and re-assembling of "un-cooperative" urban commons (Bakker, 2007: 447), marked by high initial investment costs, significant state involvement, long payback periods and risks, have resulted in disappointing results in terms of additional investments (Bayliss, 2014). A series of case studies in countries such as Spain, Argentina, Chile and the UK, among others, have pointed out the complexities and the "muddy waters of financialization" (March and Purcell, 2014: 17). Interestingly, research on infrastructure finance and financialization has also shown methodological diversity, from political economy approaches to theoretical perspectives influenced by more contingent "mezzanine-level conceptualizations", based on analysis regarding the interactions between entrepreneurial, organizational and regulatory restructuring (O'Neill, 2018). Although there is not much research on infrastructure financialization in the Global South, recent work has been undertaken on infrastructure in general (see Hildyard, 2016) or on specific sectors such as basic sanitation (Bakker, 2007; Britto and Rezende, 2017). Likewise, the number of case studies on infrastructure financialization in specific cities in the Global South is increasing (see, for example Klink et al. (2019) on metropolitan São Paulo, Brazil).

A final theme in the literature on cities in times of financialization is how neoliberalization and austerity have both renewed and made the entanglements between cities and capital markets increasingly complex. This has generated work on heterodox instruments of municipal finance such as tax increment finance districts (Weber, 2010); the management of risks in concessions, private-public partnerships (Ashton et al., 2016) and municipal operations in the capital markets; the disciplinary, post-political effects of capital markets on city governance (Peck and Whiteside, 2016); and the increasingly speculative operating of credit markets by municipalities (Kirkpatrick, 2016). Although in Brazil local and state governments are restricted by national regulation to access capital markets, recent research has investigated how municipalities have creatively bypassed this national framework in order to mobilize credit under non-transparent and contradictory conditions (Canettieri, 2017).

In short, there is no denying that the literature has recognized "financialization in cities" as a research object. At the same time, however, the work undertaken until recently reveals a built-in and unresolved tension between political economy-inspired approaches and social studies of finance-oriented work on the relation between finance and urban studies. The former provides a scenario of neoliberal urbanism and finance-dominated capitalism structured around the penetration of finance capital in the "city as a receptacle", leaving it little or no local leverage, whereby "the city is overrun by finance" (Theurillat and Vera-Büchel, 2016: 1513). The latter is framed around more contingent and open-ended trajectories, according to which cities have the capacity to invent their own praxis and spaces of financialization whereby "the landing of capital" effectively becomes a "negotiated urban anchoring of finance capital" (Theurillat and Vera-Büchel, 2016).

In this book we argue that investigating what exactly it is that defines finance in cities, as well as how it is socially constituted in city space in the first place – by state and non-state actors alike – provides valuable insights into research on what finance capital does in cities (i.e. transforming them into bundles of liquid "tradable income yielding assets") (Guironnet and Halbert, 2015). After all, considering the technological indivisibilities, long payback periods and political contestation that accompany the planning and finance of land and infrastructure networks, as well as the organization of pools of metropolitan labour (linked through the tax system to the public budget), there is nothing inherent to the transformation of city space into investment objects (Gotham, 2009). Before finance can penetrate into the reproduction of daily life in cities, it has to be socially constituted through the multi-scalar entanglements between planners, state officials, regulatory agencies, investment bankers and other intermediaries in the making of urban financialization.

There is another value added to stretching the research agenda to the emerging financialization of the "urban condition as such" (and in the end, life itself) instead of focussing on financialization *in* cities only. The latter approach has somewhat pragmatically adopted a conceptualization of the city "as a container term to include studies of urban governance, housing, real estate and the built environment" (Aalbers, 2019: 2), and triggered work on *what* is being financialized (more specifically: land, housing, urban infrastructure and environmental resources), *how* this is being done (neoliberal governance, restructuring the city-capital market nexus, performativity of economic models and devices and so on), but leaving an open question regarding *why* this process unfolds in the first place. However, as classical critical urban studies on the geography and history of capitalist cities have emphasized earlier on (Soja, 2001; Harvey, 1982; Lefebvre, 2001, Castells, 1977), cities are no containers but represent privileged, albeit contradictory, spaces for the generation, extraction, circulation and (collective) consumption of value in a continuously rescaling and restructuring of states and markets (Clarke and Bradford, 1998; Brenner, 2004). Cities also concentrate on huge social deficits and enormous economic opportunities. Therefore, increasing our understanding of the contemporary urban experience in times of re-emerging global finance requires a perspective that goes beyond financialization in cities.

In this book we develop such an approach. It is based on an investigation of the multiple and variegated entanglements between financialization of cities, on the one hand, and the "urbanization of finance" on the other. We will do so by means of a dialectic moving back and forth between political economy (analyzing what finance does to cities in terms of "extracting value without producing" (Lapavitsas, 2013)) and social studies of finance (investigating how finance is constituted and performed in cities [i.e. the urbanization of finance] in order to provide its "relational" role and innovative financial products). This will also allow us a context-specific and situated understanding of the generation and extraction of rents, value and financial profits, as well as how these processes are connected to the variegated rescaling and restructuring of cities in the contemporary finance-driven global economy.

***

A third contribution is the book's detailed focus on the geographical and historical trajectory of urban financialization in a country from the Global South. This is not only worthwhile so as to flesh out how this specific experience is different from "finance as we know it" in industrialized countries. Perhaps even more important is to see how such a focus contributes to a research agenda whereby thick case studies from the Global South provide a better understanding of the "phenomenon" of financialization itself on the global scale. The latter can be seen in light of the significant fiscal and monetary policy interdependencies and institutional-regulatory entanglements among countries, as well as the international mobility of policies and models, at multiples scales, which have become an inherent part of the shaping of the contemporary world economy and "the production of cities through global-relational connections" (Baker and Temenos, 2015: 825).

To be clear, this implies more than working out the usual claims that the geographical and historical development trajectory of the Global South is different and therefore requires breaking out of existing patterns of trade and finance. This has been the well-known and still-influential approach of the United Nations Economic Commission for Latin America and the Caribbean (ECLAC), structured around the dependency theory notions of centre-periphery and the Global South's continuously deteriorating terms of trade in global commerce. Instead, our starting point is that financialization should be grounded within an understanding according to which cities and countries of the Global South are key players in holding together the "interdependent" (albeit asymmetric) nature of the international financial monetary system. Their massive and costly holdings of international dollar reserves and deflationary management of international capital flows, as well as the contradictory impacts of these policies on domestic socio-institutional and spatial development trajectories, represent only one of the more visible dimensions of this interdependency. Thus, instead of prioritizing the Anglo-Saxon and related contexts, the entrance point here will be the investigation of the multi-scalar "making of" financialization in cities and countries of the Global South as a first step to come to grips with a better understanding of what finance is and how it is reproduced globally (Aglietta, 2018).

Such an analytical lens also allows us to contribute to Peck's (2018) call for building more productive and promising linkages between a political economy/critical geography perspective (focussed on patterns of "asymmetrical" financialization in the South within an interdependent and unstable global system of monetary governance), on the one hand, and post-structural/post-colonial urban studies, on the other (Robinson, 2011, 2015; Brandão et al., 2018). To be specific, Peck claims that post-structural and postcolonial urban studies have generated valuable contributions to innovative theory by dismantling an essentialist and universalizing narrative, structured around the western city, which has circulated during much of the twentieth century (Healey, 2013). At the same time, however, he identifies the risk of an emerging new divide in urban theory. The latter would arise from a growing distance between efforts that look for broader, structural patterns and generalizability, as opposed to theory-building embedded in the uniqueness and non-comparability of the urban experience, which altogether "appears to make a dialogue between positions and perspectives more, and not less challenging" (Peck, 2018: 171).

We will develop the story line of this book on the basis of the Brazilian setting. Moreover, as far as that is possible and relevant, we establish links with international debates and experiences. The choice of Brazil is motivated by two reasons.

First, a book on urban financialization requires "raw material", in our case, cities and what Friedman (2005) called a planning culture embedded within a national framework.[1] In that sense, Brazil provides an excellent working object considering its development trajectory since the 1930s, driven by state-mediated industrialization and urbanization, which has consolidated a relatively dense national network of cities of different sizes. Moreover, since its re-democratization in the 1980s, the country has gradually built up an internationally acclaimed track record of progressive urban reform and an associated planning culture (Klink and Denaldi, 2015). This was initially triggered, in the midst of macroeconomic crisis and restructuring that characterized the 1980s, by small-scale slum upgrading strategies driven by progressive mayors, community pressure and "insurgent citizenship" (Holston, 2009) structured around the demand for well-located and affordable land, housing and infrastructure services as basic rights. It was followed, during the 1990s, by a gradual recognition and incorporation of these local approaches into national framework policies. Likewise, the working out of a constitutional chapter on urban reform ultimately culminated in the 2001 approval of a national law called the City Statute. The latter consolidated regulatory-institutional strengthening, which empowered cities, at least theoretically, in collaborative planning efforts framed around redistributive land-use instruments, the reduction of real estate speculation and the provision of affordable, well-located and serviced land for low-income housing.[2] During 2003–2015, the rise to presidency of successive Worker's Party administrations (*Partido dos Trabalhadores*, PT) consolidated a social developmental regime of "growth-with-inclusion-without-reforms". It combined a booming economy, abundant liquidity and the implementation of a series of large-scale subsidized flagship programs for slum upgrading (i.e. *Programa de Aceleração do Crescimento*, PAC, meaning the "National Growth Acceleration Program") and low-income housing (i.e. *Minha Casa Minha Vida*, MCMV, meaning the program "My House, My Life"), which were financed by the state and built by developers and contractors (Klink and Denaldi, 2014).

Nevertheless, even in the Brazilian literature, the rise and spectacular fall of the Brazilian social-developmental regime during 2003–2015 has not been embedded within a clear analytical reading of its contradictory urban development trajectory and planning culture.[3,4] To be specific, the demise of social-developmentalism was reflected in the economic and political crisis of the country after 2013, the impeachment of former president Dilma Rouseff in 2015 and imprisonment of former president Luiz Inácio Lula da Silva (Lula) in 2018. Nevertheless, this demise could not be dissociated from the outright disappointing results of Brazilian urban reform itself. The latter had become painfully clear during the proliferation of massive street riots and manifestations in most metropolitan areas in June 2013, triggered by complaints as well as demands for something better (Harvey, 2014). These demands were aggravated by the poor conditions of public transportation and excessively high tariffs, compounded by sprawling land use, real estate speculation and the lack of decent, affordable and well-located housing (Klink and Denaldi, 2015; Amore, 2013). Moreover, beneath the surface, the agenda of urban reform itself as well as its "toolkit" of progressive-redistributive land-use instruments had been gradually filled in by developers, contractors and construction firms with active support at multiple scales from the social-developmental state itself (Maricato, 2011; Souza et al., 2019). Thus, telling how this story has unfolded not only provides an "inside zoom" on a particularly contradictory urban trajectory in times of emerging financialization but also sheds light on the somewhat under-researched role of planning and management of cities in explaining the demise of the social-developmental regimes in the Global South itself.

There is a second reason as to why choosing Brazil is particularly adequate for a book with this scope and research method. More specifically, for decades the country has been plagued by a lack of confidence in its own money and currency, as was reflected by chronically high and unstable inflation rates. This has generated enormous challenges to design and consolidate long-term capital markets that would provide credit in order to finance investments, economic growth and the expansion of housing and urban infrastructure. If, as argued by Aglietta (2018: 1), the massive rescue package articulated by the central banks during the Global Financial Crisis (GFC) in mid-September 2008 signalled that "finance and the western economy were saved by money", it also implies that money itself, "as the foundation of value and social belonging" (Aglietta, 2018: 18–80) is crucial to the constitution of financial and capital markets. In that sense, Brazil represents a prime case study in terms of understanding the contradictory rolling back and out of the development state since the mid-1980s and its entanglements with the contradictory insertion of the country into the global monetary order of Post-Bretton Woods. These movements were reflected in initial crisis and austerity measures structured around monetary stabilization, inflation control and the creation of a new currency in 1994 (the "real"), the subsequent state support for the gradual strengthening and deepening of financial and capital markets (Paulani, 2011), as well as the more recent emergence of "financialization of and through the social-developmental state" (Aalbers, 2017) and its contradictory impacts on city spaces.

This book has been written on the basis of a research project that combined a theoretical investigation into urban financialization and its relations with the (re)production of urban space in Brazil, including detailed fieldwork that has been undertaken in metropolitan São Paulo. The latter involved documentary research of files from public and private institutions, collection and systematization of secondary data sets and semi-structured interviews undertaken with planners and government officials, developers, contractors, investment bankers and consultants, among other stakeholders. Part of this work has received financial support from the National Ministry of Science and Technology.[5]

This introduction is followed by seven chapters and a conclusion.

In the first chapter, "A conceptual primer on (urban) financialization", we review the main theoretical approaches to financialization, its linkages with urban studies and planning, as well as the critique that has been raised to the rapidly growing literature. This includes a discussion on the potential for developing a research agenda on urban financialization that takes this critique seriously.

In the second chapter, "State spaces, financialization of cities and the urbanization of finance in the Global South", we then flesh out the theoretical and empirical contours of such a research program structured around a more articulate linkage between political economy-inspired approaches and social studies of finance in the investigation of the making of financialization, with a central role for cities as privileged spaces for the generation, extraction and circulation of value. The chapter is concluded with an analysis of the relevance of such an approach for shedding light on the Brazilian trajectory.

In Chapter 3, "State spaces, urban rents and finance in the Brazilian development trajectory", we analyze the trajectory of the Brazilian developmental state since the 1960s, with an emphasis on the relations between money, credit and finance and the contradictory (re)production of metropolitan spaces. For readers not familiarized with the Brazilian scenario, it provides a necessary starting point for the case studies on urban financialization in subsequent chapters. In relation to the existing critical literature on Brazilian urban studies, the chapter provides a better articulation between political economy (that tends to ignore or marginalize cities) and the planning literature (rather distant from more detailed discussions on money, credit and finance).

The following four chapters flesh out empirical evidence of the unfolding financialization process, with an emphasis on metropolitan São Paulo. The latter is Brazil's biggest and most dynamic city and the financial capital of the country. It has accumulated more experience than other Brazilian cities in terms of the articulation between its urban planning and management framework and financial markets and therefore represents a logical choice for the purposes of this book.

Chapter 4, "Planning, projects and profitability: The role of financial and institutional devices in the urbanization of finance", which is an extended version of the paper that was published earlier on with Laisa Stroher (Klink and Stroher, 2017), analyzes a specific Brazilian planning instrument as a potential constituent space of urban financialization. More specifically, the sale of securitized building right certificates (CEPACs) in large Urban (Re)development Projects (UDPs) through so-called Urban Partnership Operations, has proliferated rapidly in several cities such as São Paulo, Rio de Janeiro, Belo Horizonte and Curitiba, among others. The chapter explores how planners, financial consultants, contractors and builders/developers effectively shape the contradictory financial and physical design and implementation of UDPs with CEPACs. The initial evidence shows that there are still clear limits to the penetration of finance into city space through CEPACs. Nevertheless, continuing market-enabling regulatory rollout, the capacity of cities such as São Paulo to constantly "innovate" with similar financial-institutional devices, as well as the increasing activism of foreign investors in buying up depreciated assets of national contractors in light of the recent proliferation of corruption scandals, suggest an open-ended trajectory with a potentially more significant role for financial actors in the design and implementation of Brazilian UDPs.

Chapter 5, "On contested water governance and the making of urban financialization: Exploring the case of metropolitan São Paulo, Brazil", explores the imbricated relations between governance failures and public sector calculative practices in the emerging financialization of infrastructure networks in general and basic sanitation in particular. The specific analysis focuses on how the entanglements between the accumulation of intergovernmental debt, conflicting pricing, capitalization and valuation practices that involve state and municipal water utilities, regulatory and monitoring agencies and consultancies, are key elements in understanding the gradual hollowing out of a system of shared state-municipal water governance and its state-mediated transformation into shareholder governance. The argument is worked out on the basis of a detailed case study on the institutional and financial "battlefield" between listed majority state-owned mixed-capital water company known as, Company for Basic Sanitation of the State of São Paulo (SABESP), and municipal utility, Environmental Sanitation Services of the City of Santo André (SEMASA), a mid-sized city on the fringe of Greater São Paulo.

Chapter 6, "Building within the limits? A closer look at the My House, My Life (MHML) Program in the outskirts of metropolitan São Paulo", presents an analytical reading of the limits and potentials of this federal government's flagship social housing program in times of increasing articulations between (state-driven) finance and real estate, including a detailed case study on its impact in the Greater ABC region during the period 2009–2014. The latter is a region of seven cities with approximately 3 million inhabitants in the "industrial heartland" of Greater São Paulo (Rodríguez-Pose et al., 2001), which, until the 1970s, concentrated a significant part of the national car industry. Its leadership from labour and social movements and some of its progressive mayors also performed a key role in the gradual emergence of Holston's (2009) "insurgent citizenship" and the transformative praxis in slums and informal settlements since the 1980s. In addition to the detailed discussion on MHML itself, the chapter is concluded with a number of insights regarding the entanglements between city planning and management of land markets, the national design of finance and subsidy guideline in social housing programs and the effective outreach of redistributive urban reform.

Chapter 7, "On reframing the capital market-austerity nexus and the emerging governance of urban securitization", provides an analytical and empirical overview of how the re-emergence of austerity policies has triggered a series of "creative" municipal and state strategies aimed at accessing capital markets in spite of the prevailing Brazilian federal framework that puts restrictions on subnational market operations. After a brief overview of the international literature on crises, austerity and capital markets and its relevance for understanding the Brazilian scenario, a detailed analysis is provided on the emergence and circulation of financial-institutional devices and governance arrangements structured around the creation of "financially independent", state-controlled city and provincial securitization companies. The chapter is concluded with suggestions for further research on the entanglements between crisis narratives and the reframing, restructuring and rescaling of the nexus between urban austerity and capital markets.

A concluding chapter articulates the Brazilian experience with broader implications for a research agenda on urban financialization with political relevance for the Global South.

Notes

1 Friedmann (2005: 184) defines planning culture "as the ways, both formal and informal, that spatial planning in a given multi-national region, country or city is conceived, institutionalized, and enacted".
2 For an analysis of the City Statute in terms of its rationale, objectives and main clauses see Carvalho and Rossbach (2010).
3 To be clear, the Brazilian literature has generated several explanations for the rise and fall of the social-developmental regime during the period that the Workers Party (*Partido dos Trabalhadores*, PT) was in command of the federal administration. Most of them are structured around wrong macroeconomic choices made after 2010 – such as overvalued exchange rates, excessively high interest rates, the particular design of austerity measures which crowded out public investments and so on. (Bresser-Pereira, 2016; Carvalho, 2018) – as well as the weak and unreliable political support received from the elites in finance and industry (Singer, 2015). The blind spot in this literature, however, is the lack of attention to the intrinsically spatial dimensions of the social-developmental regime, which targeted cities and metropolitan regions as privileged arenas, particularly through its multi-billion dollar programs My House My Life (*Minha Casa Minha Vida*) and the National Growth Acceleration Program (*Programa de Aceleração do Crescimento*).
4 In a span of a few years, the widely diverging headlines of the financial newspaper, *The Economist*, managed to express well its perplexity with the Brazilian Development trajectory: "Brazil takes off. Now the risk for Latin America's big success story is hubris". Print Edition, Nov. 12th, 2009, followed by "Has Brazil blown it? A stagnant economy, a bloated state and mass protests mean Dilma Rouseff must change course." Print Edition, September 27th, 2013.
5 More particularly, through National Research Council grant no. 307069/2017–6.

References

Aalbers MB (2015) The potential for financialization. *Dialogues in Human Geography* 5(2): 214–219.

Aalbers MB (2017) The variegated financialization of housing. *International Journal of Urban and Regional Research* 41(4): 542–554.

Aalbers MB (2019) Financial geographies of real estate and the city: A literature review. In: *Financial geography working paper series*. Leuven: KU Leuven, University of Leuven, January.

Aglietta M (2018) *Money: 5000 years of debt and power*. Translated by David Broder. London and New York: Verso.

Amore CS (2013) *Entre o nó e o fato consumado, o lugar dos pobres na cidade. Um estudo sobre as ZEIS e os Impasses da Reforma Urbana na atualidade*. PhD Thesis. University of São Paulo, Faculty of Architecture and Urbanism, São Paulo.

Ashton P; Doussard M and Weber R (2016) Reconstituting the state: City powers and exposures in Chicago's infrastructure leases. *Urban Studies* 53(7): 1384–1400.

Aveline-Dubach N (2016) Introduction to the special issue: Land and real estate development in the Greater China. *Issues & Studies: A Social Science Quarterly on China, Taiwan and East Asian Affairs* 52(4): 1602001 1–11.

Baker T and Temenos C (2015) Urban policy mobilities research: Introduction to a debate. *International Journal of Urban and Regional Research* 39(4): 824–827.

Bakker K (2007) The "commons" versus the "commodity": Alter-globalization, anti-privatization and the human right to water in the global South. *Antipode* 39(3): 430–455.

Bayliss K (2014) The financialization of water. *Review of Radical Political Economics* 46(3): 292–307.

Berndt C and Boeckler M (2009) Geographies of circulation and exchange: Construction of markets. *Progress in Human Geography* 33: 535–551.

Brandão CA; Fernández VR and Ribeiro LCQ de (2018) *Escalas espaciais, reescalonamentos e estatalidades: lições e desafios para América Latina*. Rio de Janeiro: Observatório das Metrópoles e Letra Capital.

Brenner N (2000) The urban question as a scale question: Reflections on Henri Lefebvre's urban theory and the politics of scale. *International Journal of Urban and Regional Research* 24(2): 361–378.

Brenner N (2004) *New state spaces: Urban governance and the rescaling of statehood*. Oxford: Oxford University Press.

Bresser-Pereira LZ (2016) Reflexões sobre o Novo Desenvolvimentismo e o Desenvolvimentismo Clássico. *Revista de Economia Política* 36(2): 237–265.

Britto AL and Rezende SC (2017) A política pública para os serviços urbanos de abastecimento de água e esgotamento sanitário no Brasil: financeirização, mercantilizacão e perspectivas de resistência. *Cadernos Metrópole* 19(39): 557–581.

Callon M (1998) Introduction: The embeddedness of economic markets in economics. In: Callon M (ed.) *The law of markets*. Oxford: Blackwell.

Canettieri T (2017) A produção capitalista do espaço e a gestão empresarial da política urbana: o caso da PBH Ativos S/A. *Revista brasileira de estudos urbanos e regionais* 19(3): 513–529.

Carvalho CS and Rossbach A (orgs.) (2010) *The city statute of Brazil: A commentary*. São Paulo: Ministry of Cities and Cities Alliance.

Carvalho L (2018) *Valsa brasileira. Do boom ao caos econômico*. São Paulo: Todavia.

Castells M (1977) *The urban question*. London: Edward Arnold.

Charnock G; Purcell TF and Ribera-Fumaz R (2014) City of rent the limits to the Barcelona model of urban competitiveness. *International Journal of Urban and Regional Research* 38(1): 198–217.

Chesnais F (1997) *La mondialisation financière*. Paris: Syros.

Chiapello E (2015) Financialisation of valuation. *Hum Studies* 38: 13–35.

Christophers B (2014) From Marx to market and back again: Performing the economy. *Antipode* 57(1): 12–20.

Christophers B (2015) From financialization to finance: For "de-financialization". *Dialogues in Human Geography* 5(2): 229–232.

Christophers B (2017) The state and financialisation of public land in the United Kingdom. *Antipode* 49(1): 62–85.

Clarke DB and Bradford MG (1998) Public and private consumption and the city. *Urban Studies* 35(5–6): 865–888.

Erturk I; Froud J; Johal S; Leaver A and Williams K (2007) The democratization of finance? Promises, outcomes and conditions. *Review of International Political Economy* 14(4): 553–575.

Fama EF (1980) Agency problems and the theory of the firm. *The Journal of Political Economy* 88(2): 288–302.

Fields D (2014) Contesting the financialization of urban space: Community organizations and the struggle to preserve affordable rental housing in New York City. *Journal of Urban Affairs* 37(2): 144–165.

French S; Leyshon A and Wainwright T (2011) Financializing space, spacing financialization. *Progress in Human Geography* 35(6): 798–819.

Friedmann J (2005) Globalization and the emerging culture of planning. *Progress in Planning* 64: 183–234.

Froud J; Johal S; Leaver A and Williams K (2006) *Financialization and strategy: Narratives and numbers.* London: Routledge.

Froud J; Johal S and Williams K (2002) Financialization and the coupon pool. *Capital and Class* 78: 119–152.

Gotham KF (2009) Creating liquidity out of spatial fixity: The secondary circuit of capital and the subprime mortgage crisis. *International Journal of Urban and Regional Research* 33(2): 355–371.

Graham S and Marvin S (2001) *Splintering urbanism: Networked infrastructures, technological mobilities and the urban condition.* London and New York: Routledge.

Guironnet A and Halbert L (2015) *Urban development projects, financial markets, and investors: A research note.* LATTS (Laboratoire, Techniques, Territoires et Sociétés). Chairville: École des Ponts Paritech.

Haila A (1988) Land as a financial asset: The theory of urban rent as a mirror of economic transformation. *Antipode* 20(2): 79–100.

Haila A (2016) *Urban land rent: Singapore as a property state.* West Sussex: Wiley Blackwell.

Hall S (2010) Geographies of money and finance I: Cultural economy, politics and place. *Progress in Human Geography* 35(2): 234–245.

Harvey D (1982) *The limits to capital.* Oxford: Blackwell.

Harvey D (2014) *Cidades Rebeldes.* São Paulo: Martins Fontes.

Healey P (2013) Circuits of knowledge and techniques: The transnational flow of planning ideas and practices. *International Journal of Urban and Regional Research* 37(5): 1510–1526.

Hildyard N (2016) *Licensed larceny.* Manchester: Manchester University Press.

Hilferding R (1981 [1910]) *Finance capital.* London: Routledge & Kegan Paul.

Holston J (2009) Insurgent citizenship in an era of global urban peripheries. *City & Society* 21(2): 234–267.

Kaika M and Rugierro L (2016) Land financialization as a "lived" process: The transformation of Milan's Bicocca by Pirelli. *European Urban and Regional Studies* 23(1): 3–22.

Kirkpatrick LO (2016) The new urban fiscal crisis: Finance, democracy, and municipal debt. *Politics and Society* 44(1): 45–80.

Klink J and Denaldi R (2014) On financialization and state spatial fixes in Brazil: A geographical and historical interpretation of the housing program My House My Life. *Habitat International* 44: 220–226.

Klink J and Denaldi R (2015) On urban reform, rights and planning challenges in the Brazilian metropolis. *Planning Theory* 15(4): 402–417.

Klink J; Empinotti VL and Aversa M (2019) On contested water governance and the making of urban financialization: Exploring the case of metropolitan São Paulo. *Urban Studies*, forthcoming.

Klink J and Stroher L (2017) The making of urban financialization? An exploration of Brazilian urban partnership operations with building certificates. *Land Use Policy* 69: 519–528.

Lapavitsas C (2013) *Profiting without producing: How finance exploits all of us*. London and New York: Verso.

Lefebvre H (2001) *O Direito à cidade*. Translated by Rubens Eduardo Frias. São Paulo: Centauro.

MacKenzie D (2006) *An engine, not a camera: How financial models shape markets*. Cambridge, MA: The MIT Press.

March H and Purcell T (2014) The muddy waters of financialisation and new accumulation strategies in the global water industry: The case of AGBAR. *Geoforum* 53: 11–20.

Maricato E (2011) *Impasses da política urbana no Brasil*. São Paulo: Vozes.

Martin R (2002) *Financialization of daily life*. Philadelphia: Temple University Press.

Minsky H (1996) *Uncertainly and the institutional structure of capitalist economies*. Working Paper 155. The Levy Economics Institute of Bard College, Annadale-on-Hudson, New York.

Munro M and Smith SJ (2008) Calculated affection? Charting the complex economy of home purchase. *Housing Studies* 23(2): 349–367.

O'Neill P (2018) The financialisation of urban infrastructure: A framework of analysis. *Urban Studies*. Online First. Available at: https://doi.org/10.1177%2F0042098017751983 (Accessed 21 September 2018).

Paulani LM (2011) A inserção da economia brasileira no cenário mundial: uma reflexão sobre o papel do Estado e sobre a situação atual real à luz da história. In: *Logros e Retos del Brasil Contemporâneo*. Cidade de México, México, 24 a 26 de Agosto de 2011. UNAM.

Peck J (2018) Novas direções na teoria urbana: para além da comparação? In: Brandão CA; Fernández VR and Ribeiro LCQ de (eds.) *Escalas Espaciais, Reescalonamentos e Estatalidades: lições e desafios para América Latina*. Rio de Janeiro: Letra Capital & Observatório das Metrópoles, pp. 167–222.

Peck J and Whiteside H (2016) Financializing Detroit. *Economic Geography* 92(3): 235–268.

Polanyi K (2001) *The great transformation: The political and economic origins of our times*. 2nd Edition. Boston: Beacon Press.

Pryke M and Allen J (2017) Financialising urban water infrastructure: Extracting local value, distributing value globally. *Urban Studies* 56(7): 1326–1346.

Robinson J (2011) Cities in a world of cities: The comparative gesture. *International Journal of Urban and Regional Research* 35(1): 1–23.

Robinson J (2015) "Arriving at" urban policies: The topological spaces of urban policy mobility. *International Journal of Urban and Regional Research* 39(4): 831–834.

Rodríguez-Pose A; Tomaney J and Klink J (2001) Local empowerment through economic restructuring in Brazil: The case of the greater ABC region. *Geoforum* 32: 459–469.

Sanfelici D (2013) Financeirização e a produção do espaço urbano no Brasil: uma contribuição ao debate. *Eure* 39(118): 27–46.

Savini F and Aalbers M (2016) The de-contextualisation of land use planning through financialisation: Urban redevelopment in Milan. *European Urban and Regional Studies* 23(4): 878–894.

Singer A (2015) Cutucando onças com varas curtas. *Novos Estudos CEBRAP* 102: 39–67.

Soja EW (2001) *Postmetropolis: Critical studies of cities and regions*. Malden, MA, USA; Oxford, UK; Melbourne, Australia and Berlin: Blackwell.

Souza VC de; Klink J and Denaldi R (2019) Planejamento reformista-progressista, instrumentos urbanísticos e a (re)produção do espaço em tempo de neoliberalização. Uma exploração a partir do caso de São Bernardo do Campo (São Paulo). *Eure*. Forthcoming.

Swyngedouw E; Moulaert F and Rodriguez A (2002) Neoliberalization in Europe: Large scale urban development projects and the new urban policy. *Antipode* 34(3): 542–577.

Theurillat T and Vera-Büchel N (2016) Commentary: From capital lending to urban anchoring: The negotiated city. *Urban Studies* 53(7): 1509–1518.

Torrance MI (2008) Forging glocal governance? Urban infrastructures as networked financial products. *International Journal of Urban and Regional Research* 32(1): 1–21.

Weber R (2010) Selling city futures: The financialization of urban redevelopment policy. *Economic Geography* 86(3): 251–274.

Zwan vd N (2014) Making sense of financialization. *Socio-Economic Review* 12: 99–129.

1 A conceptual primer on (urban) financialization

1.1 Financialization: Main theoretical strands and research approaches

1.1.1 Introduction

Perhaps surprisingly, this chapter will not deal extensively with mainstream economics, neither its macroeconomic strand structured around concepts of rational expectations and perfect foresight (Blanchard, 1980), nor neoclassical microeconomics based on individual rational choice, decentralized functioning of prices and the emergence of spontaneous market equilibrium. Mainstream economics has, by and large, been rather silent on financialization. Particularly after the demise of Bretton Woods and the increasing influence of finance and capital markets on the functioning of economies and societies, one would have expected the issue to feature prominently on the agenda. There are three dimensions of mainstream economics that help us to understand this paradox.

The first has to do with the role of money in the coordination of transactions in self-regulating markets. While money represents the central nervous system that frames the transactions of goods and services and the circulation of value in markets, mainstream economics has never been able to come to grips with its peculiar nature (Paulani, 1991). Economics has dealt with money in a cumbersome manner within its overall framework, which is based on general equilibrium and utility maximizing consumers and producers that face given prices. In such a system, market clearing will always occur, meaning that in "n" markets for goods and services the individual marginal utility will equal prices. However, the framework is unable to account for the utility of the "n+1", the commodity represented by money, which nevertheless provides all the functional benefits for market participants in terms of reducing transaction costs. "Normal" economic science has never really found a way out of this epistemic dead-end. As a matter of fact, it ended up providing pseudo-solutions in contradiction with its own methodological foundations (Paulani, 1991). One of the well-known illustrations of these contradictions was the transformation of decentralized exchange into its opposite, as remembered by Aglietta (2018: 19): "under the aegis of a metaphorical entity, called by one of the founding fathers of modern microeconomics Walras as 'the auctioneer'". In a way, then, money is brought into the neoclassical system as an exogenous *numeraire* in order to deliver functionality and utility in terms of a unit of calculation, exchange and storage. As a consequence, and quite different from heterodox economics and other strands of the literature on financialization we will be discussing in this chapter, mainstream economics believes the role of money to be neutral. Neutrality of money means it doesn't effectively change the functioning and dynamics of markets for goods and services. Moreover, orthodox monetarism, as reflected in Friedman's (1968) quantity theory of money, claimed that money is exogenous to the economy and can (and, as a matter of fact, should) be controlled by governments and central banks. Instead, heterodox approaches developed an analysis on endogenous money. According to such a perspective, money is intrinsically linked to the economic development trajectory and social dynamics of communities and countries (Lavoie, 2006).[1] The starting point of this book is that money can never be neutral considering its social foundations and key role in the historical constitution of markets for finance and credit and, ultimately, society itself (Aglietta, 2018).

A second dimension of normal economic science that helps us to understand its silence on the issue of financialization is related to its peculiar treatment of time and the relation between the past, present and future. On the basis of portfolio theory and the capital asset pricing model (Black and Scholes, 1973), which became widely disseminated in the analysis of the pricing of derivatives and options in future markets, modern finance theory has elaborated a sophisticated version of dealing with risk and uncertainty. Financial economics has developed a time perspective whereby the expectations regarding future asset prices are capitalized into the present values of today. This was exactly what triggered Eugene Fama's (1970) famous thesis on the strong efficiency of financial markets. According to this concept, investors having rational expectations regarding the future make it challenging for individual actors to outsmart or beat the market. The reason is that asset prices capitalize the perfect foresight of individual market actors, which are all using the same model. In a way, this linkage between self-declared financial efficiency of markets and the "coordination of the present by the future" has consolidated what Aglietta (2018: 23) calls "finance's self-transcendence", whereby the financial market's subjective expectations regarding the future have increasingly become important in shaping the present.

As will be discussed in the next chapter, mainstream finance's perspective on time, risk, portfolio remuneration and value is problematic for at least two reasons. The first is that the assumption of perfect foresight of financial actors doesn't allow it to incorporate a perspective on systemic risk associated with the making, as well as the preservation, of financial markets in times of immanent threats of crisis, collapse and contagion. In these scenarios, it is highly unlikely that perfect foresight will prevail. The implication is that corporate finance only deals with risk during "normal times" and systemic risk in "differential times" during the making and preservation of finance is the state's business. A second, related problem is corporate finance's argument on the correlation between risk, as measured in terms of the fluctuations in returns obtained from particular investments as compared with a hypothetical average market portfolio, and the rate of interest used in the capitalization of future income streams to the present. In thin and underdeveloped capital markets, there is no obvious way to find this benchmark portfolio. Moreover, there is neither a logical reason nor empirical evidence that suggests a "moral" linkage between risk and investor's return. Instead, it could be claimed that rational investors try to maximize returns in low-risk environments.

The silence of mainstream economics on financialization also has direct implications for its analytical reading of the relations between money, finance and space. Harvey (2009), for example, discusses the World Bank's 2009 Development Report. The latter is framed around an analytical and empirical analysis "regarding macroeconomic growth and the reshaping of economic geography and regional development and urbanization in particular" (World Bank, 2009). The report is interesting considering it mobilizes mainstream's "rediscovery" of the spatial and scalar dimensions of development using conceptual pillars of density (the urban-metropolitan scale that concentrates economies of agglomeration), distance (which frames the connections between dynamic and peripheral regions in national space economies around notions of cost of transportation and the availability of infrastructure for connectivity) and division (which articulates the international division of labour and patterns of trade and investment at the global scale).[2] Nevertheless, as argued by Harvey, in relation to the essentially spatial dimension of the Global Financial Crisis, which triggered a series of "financial Katrinas" (Harvey, 2009: 1270), "the report is, unfortunately, not only lamentably silent, but deeply complicit with the kind of policies that got us into this mess".

This epistemic "blind spot" regarding the potentially destabilizing and non-neutral role of money and finance in the (re)production of space is also visible in the core of the neoclassical synthesis on the symbiosis between land and real estate markets and finance, which was undertaken by authors such as DiPasquale and Wheaton (1996). Their work is widely used in mainstream schools such as the Sloan Institute for Real Estate at the Massachusetts Institute for Technology (MIT) or the Institute for Real Estate within the University of São Paulo, among others. In a way, their analysis represents an eloquent update of the neoclassical urban research that had been set up in the 1960s by Alonso (1964), Muth (1969) and others, particularly considering the latter had not established a full-fledged analytical reading of the gradual articulation between cities and global capital markets. In the model of DiPasquele and Wheaton, the working of prices in asset markets for real estate-backed securities represent a perfect mirror of price dynamics in the markets for the built environment, which concentrate on buying and selling physical real estate products. For instance, increases in real residential rents caused by unexpected short-term demographic urban growth lead to an escalation of asset prices of real estate-backed investment funds. This subsequently triggers additional construction activity, a growing housing stock and gradually decreasing prices that eventually clear residential rental markets. In this world of "social physics", mismatches between demand and supply are only possible in the short run. These are largely due to low price elasticities of supply (that is, slow responsiveness of construction activity to price increases) that characterize land and real estate markets, which are exacerbated by excessively rigid land use regulations. Another factor that causes temporary mismatches is the unfavourable macroeconomic environment, mainly triggered by distorted policies that create high-risk and low investor yields and hurdles for the entrance of finance capital into real estate and housing. Needless to say, mainstream urban financial economics along the lines of Di Pasquale and Wheaton has not spent much effort to pause, breathe in and carefully reflect on some of the deeper causes and consequences of the Great Financial Crisis (GFC).

A third and related reason behind mainstream's silence on financialization in general and the GFC in particular is its emphasis on the self-regulating capacity of financial markets. In mainstream economics, the GFC has been attributed to exogenous failures of monitoring and enforcement agencies that have not done a good job in terms of maintaining proper standards for solvability and liquidity in housing finance and providing the right incentives to avoid informational asymmetries and moral hazards in otherwise self-regulating systems. The prevailing orthodox view is that the traditional macro-prudential supervision, as designed and implemented by public authorities and aimed at stricter rules and regulations in regard to financial intermediaries and banks, should be replaced by micro-prudential procedures whereby "various internal rating procedures such as Internal Rating Based and Rating Agencies' regular announcements about the soundness of banks, financial intermediaries (including rating agencies!) and innovated products and processes" will enable markets to incorporate short-term deviations from financial equilibrium and mismatches between demand and supply for investment funds (Ülgen, 2014: 7–8). In the literature on financialization, on the other hand, instability is an inherent part of the working of financial markets and requires state intervention in the control and monitoring of institutions in order to avoid systemic crises and contagion of aggregate macroeconomic circuits.

In what follows, therefore, we will deviate from each of these entrance points of mainstream economics (neutrality of money; time, risk and rational expectations and the self-regulatory capacity of financial markets) and discuss financialization from an interdisciplinary perspective that mobilizes several strands of political economy, critical management studies and shareholder value theory, social studies of finance and the literature on the financialization of everyday life.

1.1.2 Political economy

We should first, in a rather pragmatic manner, provide our definition of political economy. A useful starting point is Aalbers and Christophers' (2014: 2) discussion on the role of housing and capital in political economy. They work out a perspective on political economy in terms of its mutually constitutive guises of circulation of value, social relations and ideology. To be specific, value circulation is analyzed by them in terms of the generation, circulation and consumption, as well as the storage of value around housing in capitalist economies. The dimension of social relations is focussed on the intrinsically unequal access to housing, which has also generated high disparities in the accumulation of wealth. Finally, political economy's emphasis on ideology in housing is intrinsically linked to representations of space structured around the democratization of property and the social constitution of a society of homeowners. According to these authors, the state is actively involved in each of these constitutive guises of capital.

In Chapter 2, on conceptual bridges between critical urban studies, political economy and the research on financialization, we will briefly discuss Nitzan and Bichler's (2009) work on political economy and their conceptualization of capital itself. Their approach both defies concepts of physical capital in terms of stocks, as envisaged by neoclassical theory of the firm, as well as a Marxian reading of capital in terms of social relations only. According to them, capital is not only related to power; the essence of capital *is* power, which is defined in terms of the concept of differential capitalization. The concept provides potential insights regarding the relationships between power and financialization. Instead of a mere operational-technical device for calculating present values, as prevailing in mainstream corporate finance, "really existing" (differential) capitalization reflects the relative power of actors to influence expectations regarding cash flows, risk, return and cost of capital; as such, "the elementary particles of capitalization are all about power" (Nitzan and Bichler, 2009: 212).

For the specific purpose of reviewing the literature on financialization, we will adopt a broad definition of political economy structured around different strands of heterodox economics that nevertheless prioritize similar dimensions related to: the generation and creation of value (along the lines of classical political economy, traditionally more concerned with dimensions of production); unequal social relations and structures as well as power itself; the role of ideology and representation; and, finally, the overwhelming presence of crises, booms, busts and bubbles, not as temporary deviations from a steady state equilibrium but as a constitutive feature of the capitalist space economy.

This admittedly elastic definition of normal economic science's "alter ego" allows us to nevertheless discuss three different, but in our view complementary, lines of work on financialization grounded in what we label here as political economy, that is, neo-Marxian, regulationist and Post-Keynesian economics.

***

Somewhat different from other concepts such as globalization and neoliberalization, which have been received with varying degrees of scepticism, Marxian political economy has always actively engaged in debates on financialization (Lapavitsas, 2013). Moreover, for contemporary Marxian authors such as Lapavitsas (2013: 32), "the deeper character of the transformation of capitalism during the last three and more decades can be more easily captured by focussing on financialization rather than globalization". Hilferding's (1910) analysis on the role of finance capital under emerging monopoly capitalism in the early twentieth century illustrates the awareness of Marxian authors on financialization and their on-going influence on contemporary research (Paulani, 2011). Hilferding wrote at the time when the growing separation between professional management and owners in listed companies became widely disseminated. This led him to investigate the increasing dominance of finance over industrial capital and the crucial role of leverage and debt-financed mergers, acquisitions and investment strategies in generating financial profits "without producing".

Other Marxian inspired work has been developed over time. For instance, Arrighi (1994), who was influenced by Fernand Braudel (1982), analyzed the long-term, recurrent patterns of rising finance in world commerce. He studied the rise and fall of hegemonic trading powers from Genoa, the Netherlands and Britain up to the USA. According to him, in each of these cases the rise of finance could be associated with a declining hegemony in production and trade. In that sense, contemporary financialization as reflected in the GFC of 2007/08 represented a sign of a weakening position of the USA within the economic world order. Likewise, Chesnais (1996) investigated the increasing integration of national capital markets within a system that had become interconnected at the global scale. In what he called "the dictatorship of creditors" over industrial capital, the world economy had transformed into a "rentier" capitalism.

In a recent Marxian-oriented synthesis, Lapavitsas (2013) warns about overly simplistic and linear analyses focussed on the opposition between finance and industrial capital whereby the former would be directly responsible for stagnation and decline of the latter. As a matter of fact, industrial capital has been actively involved in the use of financial leverage aimed at the extraction of financial profits, discarding such binary oppositions.[3] Likewise, he acknowledges that, while Marxian thinking has built up a significant track record on crises and the role of finance in absorbing periodic surpluses that are accumulated in capitalist markets (Baran and Sweezy, 1966), this should not lead to premature conclusions whereby the profitability of finance represents the main cause of industrial decline. The latter argument also represents a recurrent theme in some of the Post-Keynesian literature on the role of the rentier. Moreover, several capital switching theories have investigated the allocation of surpluses from the primary (industry and production) to the secondary (the built environment) and tertiary circuit (associated with the reproduction of life and so on) without fleshing out the specific social, geographic and historical circumstances under which this would occur.[4]

Instead, Lapavitsas (2013) calls for research on financialization that is concerned with an investigation of the systemic changes in the relationships between financial and non-financial enterprises, banks and workers/households. To be specific, his analysis is structured around three inter-related transformations (i.e. financial disintermediation; the role of non-financial actors in open capital markets; and the emergence of new interfaces between banks and families-workers). Disintermediation refers to the fact that banks have diminished their traditional role in terms of connecting savings and demand for investment funds and derive significant amounts of their profits from commissions, margins and arbitrage associated with the buying and selling of financial products in capital markets. Related with this tendency, non-financial firms have reduced their dependency on banks and started to raise funds directly through their own operations in open markets. Finally, the retrenchment of the welfare state regime and the associated fall in real wages have led banks to intensify their direct relationships with workers and families through the design, marketing and provision of new mechanisms to access credit.[5] Thus, according to Lapavitsas (2013: 45–46), "the content of financialization becomes clear only after demonstrating the financialization of non-financial enterprises, banks, and households, subsequently considering the implications for mature capitalist economies as a whole".

***

Around the 1980s, a first generation of regulation theory became well known for its investigation of a presumed transformation of what was called the Welfare-Fordist regime into a more flexible system of accumulation and regulation. According to Boyer (2009), regulation theory was structured around the systemic and historic analysis of the entanglements between "the economic", that is, the accumulation regime (referring to the generation and distribution of value and surplus value (Eckhard et al., 2014)) and the political (i.e. the mode of regulation as reflected in the variety of institutional arrangements and regulatory practices that accompanied the organization and intervention of the state at multiples scales). The latter was fleshed out through the state's involvement in the wage relation, the structure of markets and the organization of the firm, as well as the design and implementation of rules for the coordination of monetary, fiscal and trade policies at national and international levels. In a way, the main object of the regulationist research agenda was the recurrent crisis, restructuring and transformation of development trajectories, which provided capitalism with its surprising capacity to reinvent itself. The first wave of predominantly French contributions, undertaken by authors such as Lipietz (1987) and Boyer (2000a), was focussed on explaining "the hidden origins of the golden age of growth", during the two and a half decades after World War Two, and its demise and shift to "the Twenty Miserable Years (1977 to 1997)" (Boyer, 2000b: 280).[6]

   The French regulation school was also quick to grasp the significance of structural ongoing transformations in the economic sphere in terms of the increasingly "finance-led growth regime of accumulation", which was considered to be the main driving force behind globalization itself. Boyer (2000b: 311–319), for example, argued that three dimensions characterized this emerging regime. The first was a restructuring within firms into the direction of what has been labelled as shareholder governance (for more details also see section 1.2). According to Boyer, there was more to this concept than the tendential alignment of the internal rate of return on investments with the expectations of shareholders that design their optimum portfolios in international capital markets.[7] As a matter of fact, Boyer (2000b: 311) claimed that, under a finance-led regime of accumulation, managers started to face a different set of pressures due to the requirement for financial profits, which also reshaped the capital-labour relationship itself. This was related to the fact that "competition is displaced from the product market (how to sell and supplant competitors) towards the financial market (how to get means of financing)". The second dimension of a finance-led regime was related to the wage-labour nexus. To be clear, during the Fordist-Keynesian arrangement the social wage level incorporated an indirect component of state subsidized housing and urban infrastructure, which was directly connected to aggregate demand, consumption and employment generation in the economy. Under a finance-led regime, the Keynesian wage-labour nexus would be gradually hollowed out and replaced by an arrangement whereby households and workers allocate their available savings in the acquisition of shares and participation in private pension funds. This is what Aglietta (2010) labelled as the "property-owning regime", considering that wealth had become an additional driver of consumption (in addition to wages and labour income).[8] As also acknowledged by "regulationists", there is considerable doubt as to whether wealth-based consumption is able to guarantee stability and maintain effective demand and consumption in light of the uncertainty and volatility that surround the entanglements of workers and households with capital markets. Moreover, social protection networks, through private health and pension schemes, tend to become more vulnerable as compared to the subsidized social wage-labour nexus that prevailed under the Keynesian welfare regime. A third and final dimension of the new finance-led regime that was emphasized by regulation theory referred to the increasingly complex relations between the state and the economy. This was due to the state's active and increasing involvement in the social constitution of money and the markets for credit and (public) debts. Monetary and fiscal policies under the Fordist-welfare regime had been structured around the double objective of full employment and stabilization of inflation levels, which had generated successive deficits, public borrowing and the accumulation of an increasing stock of debt. From the 1990s onwards, a trajectory marked by growing debt levels and stagnating macroeconomic growth had generated national economies that were more vulnerable to fluctuations in real interest rates and pressures from financial markets to maintain budgetary discipline and fiscal austerity. According to Boyer (2000b: 313), "as soon as a completely finance-dominated regime has imposed its logic, the stability of financial markets, and not so much the stability of prices, tends to become the ideal cardinal criterion."[9]

Considering its emphasis on the mode of regulation as an encompassing analytical category in the stabilization of recurrent crises under capitalism, there are relations between regulation theory and specific strands of institutional theory (Moulaert, 2005). In particular, regulation theory establishes dialogues with the investigation of institutions from a perspective of social relations, social cohesion and the embedded character of institutional arrangements. Examples of the latter perspective can be found in the work of the old American institutionalism and authors such as Polanyi (1997) and Granovetter (1985). In that sense, Moulaert (2005: 25, 27) situates regulation theory in a tradition of critical new institutionalism, which emphasizes the "historically, institutionally and territorially contextual analysis". The latter recognizes the cumbersome geographical-historical development trajectories marked by "high levels of exploitation, improper appropriation of resources (land and other natural resources but also urban infrastructure) and opportunist strategizing". As such, regulation theory refutes a reading of institutions embedded within "the premise of transacted equilibrium". The latter is common in new institutional economics whereby institutions emerge from a decentralized search of individuals for minimum transaction costs in uncertain environments (Coase, 1960; Williamson, 1985; North, 1990).

The recent work of Aglietta (2018) illustrates well the regulationist approach to institutions in general and monetary-financial ones in particular. He provides a detailed history of money, power and debt over the last 5.000 years, starting from the ancient empires, the Greek and Roman civilizations and arriving at the gold standard and the recurrent dilemmas of international monetary governance and the constitution of world money from the twentieth century onwards. For Aglietta (2018: 76), "money is not a human invention designed to solve the problems of barter" as proclaimed by mainstream economics. According to him, "money precedes the market", considering it is "an institution that perennially links the individual to society as a whole". He provides a detailed historical analysis on the role of cities, states, merchants and bankers, among others, in the constitution of money as societal credit, involving the gradual built up of trust and a project of community and nation-building and social belonging. This involves state-mediated management of conflicts between debtors and creditors and of recurrent monetary and financial crises that have marked economies and societies since ancient times.

***

Inspired by Keynes's *Treatise on Money* (Keynes, 1930) (rather than his General Theory), Post-Keynesian scholarship, through authors such as Kalecki (1971), Kaldor (1985), Hicks (1974) and Steindl (1952) and, more recently, Minsky (1981), Lavoie (2006) and Godley (1997), among others, has consolidated an influential alternative view on the role of money, credit and finance within modern macroeconomics. It has also contested both mainstream neo-classical and quantitative monetarist conceptions.[10]

The main argument of Post-Keynesians is that the money supply is not determined outside the economy. Consequently, it cannot be controlled as an independent and exogenous variable by states and central banks. Instead, money is an inseparable element of broader and highly complex macroeconomic processes structured around investment and income and employment generation. Post-Keynesians view money as endogenous and are interested in the development of a "monetary theory of production" (instead of a "real theory of exchange", as envisaged by mainstream economics).

In several aspects, the Post-Keynesian perspective of endogenous money turns normal economic science's reasoning upside down through reverse causation. One of the well-known examples refers to the Post-Keynesians' analytical reading of the relation between savings and investments. According to mainstream economics, domestic private savings drive investments; therefore, a lack of private savings (as reflected in a combination of domestic fiscal deficits and trade deficits) leads to a bottleneck on investment growth. For Post-Keynesians, however, "the investment market can become congested through shortage of cash [read: credit]. It can never become congested through shortage of saving" (Keynes, 1937; citation in Harcourt, 2006: 68–69). In other words, the effective supply of loans and credit depends on the availability of good projects (that is, proposals earning an internal rate of return – Keynes's marginal efficiency of capital – that exceeds the cost of capital) and credible borrowers with adequate collateral, rather than the availability of prior deposits/savings, gold or reserves. As such, "the scarcity of finance is purely based on a norm, a convention" (Lavoie, 2006: 58) as well as on the expectations regarding the projected returns that can be obtained from particular project proposals. In that sense, the subjective liquidity preference of banks, that is, the confidence they have in an uncertain future, represents an important benchmark for decision-making on providing loanable funds or rationing credit.[11]

Another example of reverse causation is the relation between the monetary base and the effective flow of fiduciary and credit-money in circulation. Endogenous money in the Post-Keynesian tradition implies that "high-powered" money (that is, reserves and currency held by monetary authorities) and bank money (money deposits) are purely demand driven and cannot be imposed by central banks. As such, credit and bank deposits are not a multiple of high-powered money (as claimed by the neoclassical economics through the formulation of the credit multiplier); instead, high-powered money is a fraction of the quantity of bank money in circulation (which is directly related to aggregate demand and the state of the overall economy).

A final example of reverse causation refers to inflation. In quantitative monetarism, an excessive money supply causes inflation. Post-Keynesians, however, argue that rising prices and production trigger an increasing stock of money. Inflation has non-monetary causes and originates in the context of unresolved distributional conflicts over the appropriation of national production and income (Kalecki, 1971). While in orthodox monetarism central banks are supposed to regulate the appropriate supply of money (which guarantees the stabilization of inflation rates), Post-Keynesians claim that central banks should set the appropriate benchmark interest rate aimed at the stabilization of the macroeconomic cycle; thus, the interest rate rather than the money supply is determined exogenously.

In addition to its views on money in economic life, Post-Keynesians have contributed significantly to developing a better understanding of the role of (irrational) expectations and uncertainty in generating inherently unstable and pro-cyclical economic development trajectories. Once again, money and credit are crucial, considering that banks tend to ration credit in stagnant and depressed economies while providing abundant funds in booming economies, thereby reinforcing an essentially pro-cyclical nature of laissez-faire market economies. In that respect, it is particularly worthwhile to recall Minsky's theory on financial instability (Minsky, 1981; Eckhard et al., 2014: 28) based on his notion of "stability breeding future instability". For Minsky, investments are both the driving force behind the generation of aggregate demand, income and employment and a potential source of instability. To be specific, investment depends on the availability and ease with which lenders provide credit. In stable economies lenders will progressively loosen their standards for risk assessment in order to benefit from a positive macroeconomic environment. In his specific terminology, lenders will gradually switch from stable hedge finance (whereby future cash flows generated by investments will pay off principle and interest), to speculative finance (according to which future project revenues will only recover interest payments but require to roll-over debt and speculate on further future price increases) and high-risk Ponzi finance (a scenario where borrowers are not even expected to recover interest payments, implying imminent financial restructuring and increasing debt levels). The proliferation of defaults and the concrete perspective of financial contagion in an unstable trajectory of debt-deflation will ultimately require large-scale interventions of governments and central banks.

1.1.3   Critical management studies, shareholder value and coupon pool capitalism

The literature on shareholder value emphasizes the investigation of capital market-driven restructuring of corporate governance and strategy structured around the objective of maximizing share prices and dividends. Although there are overlaps with political economy, the research on shareholder value and governance tends to prioritize an intermediate (meso) scale of organizations (such as big corporations) as opposed to the aggregate macroeconomic scale that is privileged by the former.

In their overview of the literature, Erturk et al. (2008) remind us that shareholder capitalism is not a new phenomenon considering the rise of the modern listed corporation since the mid-nineteenth century. This gradually disseminated a strategy of public offerings and limited liability for investors. Already in the 1920s and 1930s authors such as Tawney (1923), Berle and Means (1932) and Keynes (1930) had developed a liberal-collectivist critique on the contradictory role of the rentier-financier as a "parasite", with unjustified claims on the collective generation of income and wealth that tended to drive speculation, system instability and social inequality (Erturk et al., 2008: 45).[12]

Tawney (1923), for example, distinguished between what he called legitimate and illegitimate ("improper") property. The first was directly related to productive services, such as copyrights, patents and landed property effectively used to earn a (productive) living. The latter, however, including the ownership of corporate shares, was not associated "with the right to secure the owner the produce of his toil", but "passive property", "which is merely a right to payment from the services rendered by others, in fact a private tax" (Tawney, 1923 cited by Erturk et al., 2008: 57). The increasing separation between what he called industry – represented by managers responsible for technical efficiency and the delivery of goods and services that will be bought by consumers – and business, composed of financial management, the board of directors and shareholders, a group predominantly interested in the generation of financial profits – provoked concrete tensions within modern industry and in relation with its consumers. These tensions aggravated under shareholder-driven capitalism considering that "the link which bound profits to productive efficiency is tending to be snapped" (Tawney, 1923 cited by Erturk et al., 2008: 61).

Along the same lines, Berle and Means (1932) considered that the rise of the external shareholder in the modern corporation – substituting the previous arrangement of the insider shareholder as a "quasi-partner" – could be analyzed in terms of an exchange between liquidity and the acquisition of control in the last instance (considering that daily control had been placed under the responsibility of executive management). These authors also argued that the modern corporation should neither exclusively serve the interests of owners (shareholders), nor the objectives of control (i.e. executive management); instead, top management should work out a strategy that would mobilize community interests through fair wages and employment security, stabilization of business and the provision of quality products and services. If successful, such a strategy would inevitably lead to a scenario whereby "the interest of passive property would have to give way" (Berle and Means, 1932 cited by Erturk et al., 2008: 309–312).

Nevertheless, at the time liberal-collectivists were writing, the rentier-financier was still in a minority position, certainly when compared with the role of the shareholder in contemporary finance-driven capitalism. Share-ownership effectively proliferated during the golden years of the 50s and 60s of the Keynesian-Fordist era. It was driven by the growth of company co-funded pensions schemes, which intermediary fund managers used in order to channel investments into shares. This had impacts on corporate governance, product and capital markets and required a critical update of the older work on the role of the rentier-financier from the early twentieth century (Erturk et al., 2008). This would also provide insights on the relations between the dissemination of shareholder value, both as an ideology and effective practice in the reshaping of organizations, and the contemporary debates on financialization.

Somewhat different from the critical views on the role of the shareholder in industrial capitalism during the 1920s, the recent popularization of share-ownership has divided academic debates. On the one hand, authors from modern finance theory such as Fama (1970) and Jensen and Meckling (1976) argued that efficient capital markets provided the ideal environment to guarantee the discipline and alignment of executive management with the strategic objectives and interests of shareholders. In other words, shareholder value operationalized both the carrots (complementary remuneration through stock options) and the sticks (hostile mergers and acquisitions) which avoided potential principle-agency dilemmas and guaranteed that executive management would walk the line defined according to stockowners' interests.[13] On the other hand, Lazonick and O'Sullivan (2000) were critical of the shift to shareholder value in corporate America. In their view, large corporations had moved away from a strategy driven by "retain and reinvest" to an approach marked by "downsize and redistribute". This had generated clear negative effects on product quality and functional income distribution between capital and labour in the modern enterprise. To be clear, the growth, diversification and organizational segmentation of modern corporations had, by itself, created additional complexity and challenges for corporate management to oversee how strategies of "retain and reinvest" could be reinvented in a more competitive environment and whether their organization would have the capacity to do so. In the meantime, particularly from the 1990s onwards, deregulated financial markets triggered an explosion in top-management pay, indicating a re-alignment with the premises of shareholder value.

Froud et al. (2001) use the concept of coupon pool capitalism in order to specify and update the concept of shareholder governance to the contemporary scenario of a finance-driven economy. A massive circulation of shares (coupons), which connects savings and investments, characterizes the contemporary setting. According to them (Froud et al., 2001; Froud et al. 2006), this means a gradual transformation of the "passive coupon-clipping shareholder of the 1920s to the virtual shareholder of the 2000s represented by a professional fund manager". Coupon pool capitalism is accompanied by the spread of asset ownership among workers and households and the rise of what they call "the new working rich". The latter are high paid and fee-earning financial intermediaries acting in open markets, which emerged since the restructuring and disintermediation of banking itself. At the same time, however, Froud et al. (2006) analyze the "infelicity" (defined by them as the discrepancy between "saying and doing") and the distantiation between the narrative of shareholder value and the effective numbers. They analyze a representative sample of the 100 largest UK companies (measured through market capitalization on the London Exchange) and their 500 US counterpart companies, quoted on the New York Stock Exchange and the NASDAQ market. Their findings show that increases in shareholder returns (the sum of share price appreciation and dividends) can by and large be attributed to external influences. The latter are driven by tendencies in specific industries that have pushed up share prices (in sectors such as telecommunications and high-tech firms), which are unrelated to clever management strategies aimed at shareholder value. Likewise, more conventional accounting ratios (such as the rate of return on capital employed or ROCE) show disappointing results for shareholder value when comparing the year 1983 to 2002 (Froud et al., 2006: 80–83).[14] The authors summarize their analysis on the tension between the numbers and the narratives of shareholder value according to two related dynamics. On the one hand, there is a harsh reality of competitive product markets which generated effective challenges for management to provide "value for money" to shareholders. On the other, there is the emergence of increasingly sophisticated narratives of top executives and the media in terms of shareholder value. In a way, the approach of Froud et al. (2006) provides conceptual linkages with a literature on the (counter) performativity of economic models and the social constitution of finance. This is a third strand of literature on financialization that we will discuss in the next section.

1.1.4 *Finance and the everyday, social studies of finance and performativity*

A third strand of work on financialization prioritizes the investigation of the multiple entanglements between finance and everyday life. It provides a detailed micro perspective on how, after the demise of Bretton Woods and the hollowing out of the Keynesian welfare state and its stable wage-labour relationship, finance has penetrated into our daily life. This has occurred through the rapid dissemination of innovative mechanisms in areas such as mortgages and real estate finance, private health, pension and retirement plans, consigned credit, credit cards and micro-finance, among some of the well-known examples. The work along these lines has by and large an interdisciplinary character, receiving contributions from such disciplinary areas as cultural economics (Thrift, 2001), economic sociology (MacKenzie, 2005; Callon, 1998), science and technology studies (Latour, 2005), anthropology and critical financial geography (Hall, 2010), among some of the more common influences.

It should also be stressed that there are certain overlaps between the micro-approach from finance and the everyday life and some of the other work on financialization we have discussed in this chapter. For instance, the detailed investigation of the penetration of mortgage securitization and private defined contribution pension schemes into daily life echoes Harvey's (2013) work on capital switching and the role of the secondary (real estate and the built environment) and tertiary (research and development and the reproduction of human life) circuits in absorbing recurrent industrial crises under contemporary capitalism.

Nevertheless, the methodological entrance point of this research program is different, considering its emphasis on the individual subjective management of risk, which has become an intrinsic dimension of life under a finance-driven economy. To be clear, under the Fordist welfare regime, the management of risk was shared by the state, which absorbed a significant part of the reproduction cost of labour such as housing and urban infrastructure, health and education. In the emerging scenario, however, Martin (2002: 52) argues that:

> Finance offers a word to the wise and foolish alike: Risk . . . Financialization . . . establishes the routinization of risk. . . . To be risk averse is to have one's life managed by others, to be subject to their miscalculations, and therefore to be unaccountable to oneself.

Thus, financial self-management becomes essential to plan for an uncertain future in terms of possible unemployment, health problems and retirement. For Martin (2002), life itself has become an asset to be managed (Zwan, 2014).

The gradual transformation of ourselves into financial subjects raises additional questions for researchers working on financialization of the everyday. One of them relates to the role of financial discourse and representation, imaginaries and models that accompany the internalization of finance into our subjective decision-making. In that sense, Langley (2004) provides an interesting example of how the mobilization of specific financial narratives and representations has influenced the transformation and restructuring of Anglo-American pension systems from defined benefit (DB) to defined contribution (DC) schemes. His analysis shows that a grand narrative structured around the financial deficits that were attributed to the DB arrangement, which guaranteed payments according to the last salary received and was based on co-funding by employees, was instrumental in changing to DC schemes. The latter would gradually shift the risks and burdens of pension systems towards workers. According to Langley, the emphasis on the accumulated deficits under the prevailing system and the subsequent focus on how to provide solutions by changing the actuarial-financial formulas of member contributions sidestepped the more crucial point (i.e. how the increasingly speculative and risk-taking investment behaviour of pension funds had been able to compromise the financial sustainability of the previous system in the first place).

In that sense, as a spin-off from research on the financialization of the everyday and social studies of finance, a subsequent debate emerged on the idea of performativity. The concept of performativity originated in the sociological literature on science and technology and agency-network theory (Latour, 2005). More specifically, Callon (1998) and Mackenzie (2005) conceptualized markets in general, and financial markets in particular, as sociotechnical *agencements*. The latter could be considered "as a combination of material and technical devices, texts, algorithms, rules and human beings that shape agency and give meaning to action" (Berndt and Boeckler, 2009: 543). Their work paved the way for a more specific literature on the performativity of economics, suggesting that economic science not only describes reality but also its concepts, metrics, devices and models are actively used by public and private actors alike in the social constitution of markets. Berndt and Boeckler (2009: 543–544), on the basis of Callon, argued that "it is less that academic economists (termed 'confined economists') see to it that the 'model of the world becomes the world of the model' (Thrift, 2000: 694), but rather that the practitioners of sociotechnical economic disciplines such as accounting, supply chain management or consulting (termed 'economists in the wild'), frame and perform markets by defining standards, surveying exchange processes, benchmarking goods, calculating prices and so on". On a similar line, mainstream academic economists Faulhaber and Baumol (1988) analyzed that, just like medicine and engineering, under specific situations the metrics and technical devices invented by economists have generated innovations that have contributed to changing the world in which we live.

One of the initial well-known applications of the concept of performativity occurred in the area of corporate finance, more specifically in relation to the Black and Scholes (1973) model used to describe and predict the pricing in option and future markets. MacKenzie (2005) argued that, considering the large number of traders that were using the same model, as well as feeding it with inputs based on a series of converging norms, conventions and information, the effective trading practices and prices were being influenced by the theoretical prediction of the model itself.[15] Similar work has concluded that financial valuation practices have the potential to gradually penetrate into the domain of cultural, social and environmental policies as well. As mentioned previously, for example, Chiapello (2015) discussed how forward-looking financial cash-flow projections and models have changed traditional historical-cost accounting practices used in these policy areas.

Finally, it should be observed that there is a tension between political economy and research on performativity (Christophers, 2014a). The main point of divergence, from the point of view of political economy, is the emphasis of performativity theorists on methodological individualism and their supposed refusal to embrace more structural analysis and investigation of social relations. However, as we will argue in more detail in the next chapter, this critique doesn't recognize the variety of approaches that have emerged within the literature on performativity, including investigations embedded within an analysis of social relations and the political economy of valuation associated with material and discursive strategies of specific actors within professional communities. The idea of weak performativity goes into the same direction (Christophers, 2014a; Berndt and Boeckler, 2009; Henrikson, 2009). The latter grounds performativity within an analysis of the specific mechanisms and the politics through which narratives, norms, conventions and metrics circulate and influence the collective *agencements* of professional communities. This also opens a concrete methodological perspective structured around the idea of going back and forth from structures (political economy and the politics of valuation) to markets (performativity), and from markets to structures.[16]

1.2 Financialization in cities

Particularly after the global subprime crisis, financialization has increasingly been connected with the transformation of cities and metropolitan areas after the demise of Bretton Woods (Theurillat and Vera-Büchel, 2016; O'Neil, 2018). Although it is challenging to provide a consistent overview of the rapidly growing, diverse strands of work that have emerged over the last three decades or so, the literature has emphasized a variegated but nevertheless increasing penetration of finance capital into city spaces. The theoretical and empirical investigations have concentrated on four interrelated themes: the financialization of land and real estate; neoliberalization, urban entrepreneurialism and large urban development projects; infrastructure financialization; and the city-capital market nexus.

1.2.1 Land and real estate

A first, and perhaps more obvious entrance point, has been the link between finance, land and real estate. This has generated a vast literature on how "real estate became 'just another asset class'" (Van Loon and Aalbers, 2017). To be specific, considering its high upfront costs, there is nothing particularly new about the relation between real estate and credit markets. Credit is essential in diluting the high financial burden associated with the acquisition of real estate products. Until the 1970s, this consolidated a relatively stable and specific market niche that connected housing and real estate with providers of credit. Nevertheless, the post-Bretton Woods re-emergence of global finance and the rise of creative financial engineering, secondary markets (see Chapter 2) and related institutional innovations (Helleiner, 1994), have gradually brought land and real estate back on the radar of investment bankers. The latter are constantly looking for new asset classes that allow optimizing the risk-return relation within global portfolios. Moreover, institutional investors such as pension funds, which for decades have owned housing and real estate, have gradually shifted to strategies that delegate to global real estate funds the ownership and management of their portfolios (Aalbers, 2019: 7). These transformations have now proliferated analytical and empirical work on how land and real estate – marked by high upfront costs, long payback periods, illiquidity and risks of moral hazards associated with the provision of credit to clients that reveal insufficient or incorrect information – have gradually been reshaped into tradable interest-yielding assets within global investment portfolios. Gotham (2009: 355), for instance, provides an analysis on how the secondary mortgage markets were instrumental in constituting what he calls "liquidity out of spatial fixity". Likewise, Aalbers (2019) reviews the literature on the multiple entanglements between housing and non-residential products, governance and state strategies in the making of global investment portfolios in real estate.

As a matter of fact, already in the 1980s authors from political economy and critical geography had detected a tendency for land to become transformed into a financial asset. For instance, Harvey updated Marxian perspectives by arguing that land increasingly performed the role of a financial asset in the contemporary global economy. Land markets provided capitalists with information on yields and land rents required for the decision-making processes on the transformation of urban space (Harvey, 2013). Haila (1988) claimed that, although Harvey was right in pointing out a general tendency, there was nothing particularly immanent regarding this transformation of land. According to her, it was necessary to investigate the historical and institutional conditions under which this would effectively occur. In that sense, her detailed historical fieldwork on Singapore as a South Eastern "property state" defied simplistic binary reasoning and pointed out the country's hybrid regime of land planning and management. On the one hand was the city-state's active involvement in the successful mobilization of use values aimed at providing affordable housing to its domestic residents. At the same time, however, the country was keen on the maximization of land rents in order to reap benefits from its competitive position within the global financial economy (Haila, 2016). Similar work on the limits, potentials and contradictory effects of the state in mobilizing land and real estate as a financial asset in times of neoliberalization has been undertaken in the context of Belgium and the Netherlands (Van Loon et al., 2018), the UK (Christophers, 2017), post-GFL Ireland (Byrne, 2016), France and New York (both experiences involving the private rental system) (Wijburg, 2018; Fields, 2014) and Brazil (Klink and Denaldi, 2014), while López and Rodríguez (2011) provide a detailed geographical and historical analysis of the making of the Spanish subprime crisis during 1992–2007. Rolnik (2018) develops a global empirical perspective on the financialization of housing and real estate on the basis of her work as a special rapporteur on the Right to Decent Housing of the United Nations Systems during 2014–2018.

This general perception regarding the importance of grounding the variegated financialization of housing and real estate in the specific historical, geographical and social-institutional context also triggered research on the role of expectations, as well as the circulation of norms, conventions and calculative practices used in the pricing and valuation of land and real estate. In that sense,

Christopher's analysis on the role of viability studies in the design and implementation of social housing policies in the UK represented a powerful illustration of how real estate developers and consultants – called by Callon (1998) as "economists in the wild" – are actively involved in the "performative" use of economic and financial models in "making and remaking the urban world, as opposed to merely describing it in some passive, detached way" (Christophers, 2014b: 79). On the other hand, Crosby and Henneberry (2016) and Hennebery and Roberts (2008) investigated in detail the calculative practices of real estate firms. They only found a partial penetration of financial techniques and metrics commonly adopted by capital markets and investment bankers, while traditional pricing and valuation methods and comparative benchmarking continued to prevail. In the specific Brazilian context, Sanfelici (2013) has analyzed how São Paulo-based developers, in a macroeconomic context of growth, with abundant liquidity and positive expectations, used IPOs to build up speculative land banks in the cheaper states in order to profit from the post-2007 real estate boom, and why this eventually led to overly optimistic sales projections, excess supply and bursting bubbles.

1.2.2 Neoliberalization, urban entrepreneurialism and urban redevelopment projects

The intense involvement of the state, at multiple scales, in the mobilization of land and real estate as a financial asset has generated a related, more specific literature on the transformation of urban planning itself in times of neoliberalization and financialization. As mentioned by Guironnet and Halbert (2015), the emergence of urban entrepreneurialism through large redevelopment projects and competitive city strategies is not a new theme and has been discussed at length in the literature since Harvey (1989) and Swyngedouw et al. (2002), among others. Nevertheless, this research didn't flesh out the specific role of finance capital in the transformation of governance and planning. More recent work on the entanglements between finance capital and urban planning through large urban redevelopment projects has now tried to fill in this gap. For instance, Guironnet and Halbert (2015), Guironnet et al. (2016) and Halbert and Attuyer (2016) develop a conceptual framework and a series of case studies on the circulation of norms and conventions within professional communities, with actors in and outside finance, regarding the pricing and valuation of real estate aimed at transforming city space into "a portfolio of tradable income yielding assets" through large urban development projects. Likewise, Guy et al. (2002) argue that differences in development cultures, for example, between institutional investors and independent real estate firms, should be taken into consideration by planners in the design and implementation of urban regeneration projects. A growing empirical literature has started to trace the role of finance and intermediaries in the design and implementation of projects (Rutland, 2010). Kaika and Rugierro (2016), for example, focussed on the active role of Pirelli in transforming its industrial land into a financial asset through the involvement in urban redevelopment projects in Milan. Likewise,

Charnock et al. (2014) and Savini and Aalbers (2016), on the basis of the experiences in Barcelona (Spain) and the city of Sesto San Giovanni (located in the industrial outskirts of Greater Milan/Italy), respectively, pointed out the increasing entanglements between financialization and urban entrepreneurialism itself, as reflected in the involvement of the (local) state in mobilizing finance capital for the delivery of land, infrastructure and real estate. Specifically in relation to the transformations in the outskirts of Greater Milan, Savini and Aalbers (2016: 883) argued that planning was directly involved in an ongoing de-contextualization of land use, whereby:

> the decision-making over land use programmes (such as typology and number of properties) is less closely linked to the decision-making on investments. While the first tries to match the needs of local inhabitants to local economic circumstances, the second refers to profit expectations of shareholders.

Although the literature has predominantly focussed on the OECD region, work on the Global Urban South is increasing. Rouanet and Halbert (2016), for instance, show the interactions between municipal and state governments, global finance capital and local developers in the city of Bangalore, India. The latter manage to extract value in light of their superior leverage over tacit information networks that are essential in the design and implementation of urban development policies. Likewise, there is a growing literature on large urban redevelopment projects and finance in the Brazilian context. In particular, significant work has been done on Urban Partnership Operations (UPO), especially in the city of São Paulo. The latter has experimented extensively with this instrument since the 1990s. UPOs allow local government to design and implement redevelopment projects aimed at social, urban and economic transformation within a predefined area perimeter, whereby projected infrastructure and housing investments are financed through the sale of additional building right certificates to interested developers and investors from the capital markets (Sandroni, 2010). The experience with UPOs in São Paulo and, more recently, in Rio de Janeiro, has generated a critical review of the entanglements between the state, capital markets and planning through urban redevelopment projects in Brazil (Klink and Stroher, 2017; Mosciaro and Pereira, 2019; Fix, 2007). In Chapter 4 we will present a detailed case study on this instrument.

1.2.3 Infrastructure financialization

Brenner's (2004) work on the restructuring of spatial Keynesianism and Graham and Marvin's (2001) analysis of global infrastructure transitions provided complementary representations of how the neoliberal project and the demise of the welfare state were reshaping territorial governance in general, and the planning and operation of infrastructure networks in particular. In their view, the emergence of a localist, competitive regime of governance was instrumental in gradually reshaping the modernist infrastructure ideal of integrated networks. The latter was structured around a project of socio-spatial cohesion, nation-building and universal access to networks financed by a mix of general-purpose resources, taxes and cross-subsidized prices. This project was coordinated by public monopolies operating at the macro-regional or national scale.

Reshaping that ideal implied that infrastructure's intrinsic characteristics as a public good – marked by indivisibilities, economies of scale and non-rivalry in consumption – would have to be hollowed out gradually. Technological progress facilitated the fragmenting and unbundling of specific parts of infrastructure networks and triggered an increasing interface between international and local actors in the design and implementation of "glocal" segmentation strategies. Segmentation could unfold through territorial (e.g. prioritizing niche and high-income neighbourhoods) or functional (e.g. underinvesting in sectors where direct charges were cumbersome, such as storm-water drainage) strategies. Irrespective of the specific design, this would generate spatially and socio-economically selective and fragmented regimes of infrastructure provision. The increasing participation of private players would also mean a gradual bypass of low-income segments while "cherry-picking" the more solvent parts of the networks.

Graham and Marvin provided a powerful contribution to critical thinking on neoliberal infrastructure regimes and the associated dissemination of commodification and privatization of networks. Nevertheless, it remained unclear how infrastructure transformations were both influenced by and also affected the re-emergence of global finance. More recent literature has provided contributions to understanding the imbrications between financialization and infrastructure transformations (O'Neill, 2018).

A recurrent theme in these debates is the relational role of investment bankers and other financial intermediaries. In theory, they provide the required innovative financial engineering that enables the creation of diversified portfolios and a reduced direct exposure of investors to risk by transforming illiquid, chunky and high-cost infrastructure networks into desirable investment assets, and receiving fees for exactly doing that (Pryke and Allen, 2017; Torrance, 2008; Bresnihan, 2016). Nevertheless, the transformation of engineering objects into financial assets is not a smooth, linear process, which would imply costless unbundling, repackaging or disassembling and re-assembling of lumpy networks. As a matter of fact, infrastructure is an "uncooperative" commodity (Bakker, 2007: 447), characterized by high upfront investment costs and significant involvement of the state, including its agencies responsible for regulation and enforcement, long payback periods and high risks associated with regulatory-institutional changes.

Therefore, despite the widespread circulation of narratives on the advantages associated with increased private-sector participation in the finance, building and operation of infrastructure,[17] the available evidence is outright disappointing (Bayliss, 2014; March and Purcell, 2014; Newborne and Mason, 2012). Nevertheless, more specifically referring to water, Bakker (2013: 253) argues that a "strategic retreat of private companies from certain countries and regions", the strengthening of global campaigns structured around the human right to water (as reflected in the 2010 statement by the United Nations Human Rights Council) and the promotion and gradual re-emergence of alternative arrangements to privatization, such as public-public partnerships and re-municipalization, should not be interpreted in terms of a retreat from water liberalization. The human right to water, for instance, is firmly grounded in a liberal conception that prioritizes private ownership and individual rights, as opposed to a perspective structured around the collective right to common pool resources (Aversa et al., 2018). Moreover, a critical view on public-public partnerships has indicated contradictions in the emerging alternative arrangements in water governance. One of these is state corporatization, whereby public companies incorporate private values structured around effectiveness and efficiency, which both enable a preferential entrance of private actors into the delivery of public services and blur boundaries between the private and public domains (Rooyen and Hall, 2007; Boag and McDonald, 2010).[18]

Increasing complexity and blurring boundaries between state and non-state actors in the planning, finance, building and operation of infrastructure networks require a more detailed look at the *meso*-intermediate level. This also allows a more elaborate articulation between macro-structural transformations and concepts, such as neoliberalization and financialization, and the variegated and situated experiences of specific infrastructure transformations on the ground. In that sense, Ashton et al. (2016) analyze the contractual relations and the asymmetric risk-sharing between city governments and capital markets in light of the likely occurrence of liabilities not accounted for after the negotiation and signature of lease contracts that cover extended periods for investment, operation and maintenance of infrastructure systems. Likewise, Deruytter and Derudder (2019) discuss the entanglements between the Belgian federal state and the global Australian company Macquarie Bank in the financialization of the Brussels airport. Their analysis shows a complex arena whereby liberal market economy-inspired arrangements become intertwined with more informally coordinated premises that are characteristic of the Belgian institutional framework. Critical to political economy as an entrance point, O'Neill (2018: 1) develops in a similar way a framework on the "relationships between infrastructure investing and the infrastructure-enabled flows of a city through the lens of three "mezzanine-level conceptualizations" of "capital structure, organizational structure and regulatory structure", so that the "particularities of political economy that pervade the city and its economy become prominent as a consequence" (O'Neill, 2018: 3). In her recent overview of the literature on infrastructure financialization, Whiteside (2019) separates between what she calls issues related to infrastructure funding (that is, the cash flows and flow of funds required to deliver infrastructure) and finance (the institutional devices and governance frameworks that are required to mobilize these cash flows). Moreover, she argues that, over and above "the strategically important financialized infrastructure" (i.e. the networks that provide returns to institutional investors on the basis of specific asset classes), there has hardly been any research on finance (and financialization) of social infrastructure. This is disappointing, considering that "infrastructure is a venue for service provision, the structure of community development and urban planning, the contour of property rights, an urban growth programme concretized and an enabler of democratic decision-making" (Whiteside, 2019: 6). Such an extended approach would also provide a fresh perspective regarding the limits and potentials of finance in "soft services" such as cleaning, catering, public services, care work and, we would claim, planning itself. Moreover, her review of the literature indicates a lack of analytical and empirical research on infrastructure finance and financialization in the Global South.

An important exception to the latter point is the work by Hildyard (2016: 23) on Asia, Latin America and Africa. It stresses the interfaces between the state and finance capital in the value extraction from infrastructure networks through the design of guaranteed and contracted income streams. Likewise, Bakker (2007), in global terms, as well as Britto and Rezende (2017) and Klink et al. (2019) for the Brazilian scenario, analyze the selective entrance of finance and the private sector in water and basic sanitation.

1.2.4 The city-capital market nexus

A final theme of the literature on financialization in cities is structured around the analysis of how neoliberalization and austerity have both renewed and made the entanglements between cities and capital markets increasingly complex. This has generated work on tax increment finance districts and other non-orthodox instruments of municipal finance (Weber, 2010); the increasingly risky strategies of cities in accessing capital markets (Kirkpatrick, 2016); and the disciplinary effects of capital markets on city governance (Peck and Whiteside, 2016), among some of the more frequently discussed topics.

Tax increment finance (TIF) districts are probably among the most discussed non-traditional instruments used to finance urban development in US cities (Pacewicz, 2013). Although there is considerable differentiation in design, the original idea behind TIF was to provide local governments with an instrument for social, urban and economic transformation of derelict areas. The projected additional (property) tax resources resulting from urban redevelopment would allow local governments to raise upfront resources in municipal bond markets so as to implement the required investments. TIF districts have divided academic debates. While authors such as Brueckner (2001) argue it represents a tool to provide local public goods within areas that geographically match beneficiaries and contributors, Weber (2010, 2015) describes the instrument's contradictory impacts in terms of triggering speculative bubbles and excessive municipal debts. Pacewicz (2016), however, according to what she calls a "politics of earmarking perspective", claims that the proliferation of TIF in the institutional context of US municipal governance reflects local government's strategy aimed at augmenting discretionary resources as compared to a setting whereby a significant part of the existing tax and resource base has been earmarked for specific (non) city purposes (such as school districts).

The US institutional setting has generated other heterodox instruments that are used by city governments in times of fiscal austerity and neoliberalization. Botein and Heidkamp (2013), for example, describe how the sales of local property tax liens were designed to alleviate fiscal stress but eventually generated a continuously shrinking tax base and new impasses for the regeneration of derelict areas.

Another approach in the literature on the city-capital market nexus is the critical analysis of its built-in pro-cyclical profile towards risk. In that respect, Kirkpatrick (2016) revisits Minsky's work on financial instability and Polanyi's analysis on the fictitious nature of money, land and labour. He argues that US local governments have been increasingly driven into risky "Ponzi-esque debt structures", whereby projects do not cover amortization and interests and require additional debt emissions. This has also gradually hollowed out any social control over the activities of "de-democratized" quasi-public specific-purpose bodies, which are responsible for the management of complex municipal operations in capital markets (see also Block, 2016).

A final area of interest in the literature on cities and capital markets relates to the restructuring of urban governance itself under the influence of capital markets. Peck and Whiteside (2016) and Peck (2017a, 2017b) develop an analytical framework that provides new insights regarding the transformations that have been taking place since the 1980s from late entrepreneurial towards neoliberal and financialized forms of urban governance. Peck's empirical focus is on austerity urbanism in US cities such as Detroit and Atlantic City, where successive cycles of productive restructuring and downsizing have effectively hollowed out cities' economic and fiscal base. More particularly, these cities' increasing dependency on capital markets ("bondholder value") and the subsequent deterioration in their liquidity and payment conditions have led to continuous threats of bankruptcy. This has set the stage for the emergence of what he calls sites of "Ceasarian", post-political governance (Peck, 2017b: 353). It is characterized by a gradual takeover of city planning and management in general, and of municipal contracts, assets and personnel in particular, by emergency managers. External management is called in by the state in order to realize the financial restructuring and "rightsizing" of municipal governance. As Mayor Guardian of Atlantic City mentioned, "he was being pressured to accept a 'one sided surrender' by state authorities, apparently on a mission to 'abolish local government'" (quoted in Johnson, 2016: 1 as cited by Peck, 2017b: 353).

Although in Brazil, local and state governments are constrained by national regulation to access capital markets, recent research has analyzed the proliferation of heterodox institutional devices in order to bypass these restrictions and tap into debt finance. For instance, the city of Belo Horizonte, which is the capital of the state of Belo Horizonte (located in the south-eastern richer region of the country), has set up a specific-purpose semi-public mixed-capital company, which was responsible for the management of municipal assets (such as land, infrastructure networks and tax liens) and the mobilization of additional financial resources. Canettieri (2017) analyzes in some detail how this company became involved in complex and non-transparent negotiated sales of debentures to selected investors while using municipal assets and tax receivables as guarantees. This eventually severely compromised the financial liquidity and solvency of the city itself.[19]

1.3 The limits to financialization

The rapidly proliferating research agenda on financialization has received three general critiques.

First, in a stock-taking exercise Christopher (2015) argued that much research has lost political relevance considering it has prioritized the analysis of "what finance does" (i.e. extracting value from all of us without producing), while neglecting the investigation of "what it is" (and "how it is constituted socially"). Part of this is due to the built-in tensions between the various strands of the literature we have discussed in section 1.1. To be specific, political economy in general, and Marxian strands in particular, have traditionally focussed on broader historical processes of value generation and appropriation, while taking the distribution, circulation and consumption of value, which take place in markets, for granted. At the same time, markets have been the privileged objects of social studies of finance. There has been surprisingly little work aimed at building connections between these strands. To illustrate, in his overview of the literature, Lapavitsas (2013), working in the tradition of political economy, briefly discusses shareholder value (claiming its similarities with regulation theory) but doesn't touch upon the research that has been undertaken by scholars of social studies of finance and performativity. On the other hand, in their research on shareholder value and restructuring of corporate governance and strategy, embedded within social studies of finance and performativity, Froud et al. (2006: 130) provide a detailed analysis regarding "infelicity" and the discrepancies between narratives and numbers. As was discussed earlier, the latter are reflected in the mismatches between the stories told by the financial media and chief executives regarding the dominance of capital markets and the effective delivery of value for money for shareholders in competitive product markets. While doing so, however, the authors' verdict on the potential of political economy in providing additional insights is harsh:

> If we then ask how the different elements of the narrative and performative all fit together, the answer seems to be in unstable, unique configurations very different from those envisaged in traditional political economy, with its 'box and arrow' approach, centred on stable configurations with durable, mechanical links from institutions conduct and performance in models of capitalism.

In this book we will argue in favour of approaches that allow us to engage with questions regarding what finance is and what it does by moving dialectically "from Marx to markets and from markets back to Marx", as suggested by Christophers (2014a). In that sense, a more careful articulation between political economy and social studies of finance-inspired work on performativity can provide valuable insights to research on financialization.

A second critique refers to the fact that the prevailing literature has emphasized the investigation of "finance in cities" as enabling receptacles of financialization. However, considering them as privileged spaces for the generation, circulation and extraction of value and (collective) consumption in the contemporary world economy (Clarke and Bradford, 1998), cities are not only influenced by finance capital but are actively engaged in the making of financialization itself. As a matter of fact, before cities – as complex and relatively indivisible socio-technological networks of land, infrastructure and pools of labour – can be transformed into liquid (i.e. tradable) and interest-yielding assets (Gotham, 2009), credit and financial markets that provide commensurability in urban space have to be created in the first place. As such, the existing literature has been silent on the historically and geographically variegated entanglements between the financialization of cities and the role of national states, cities and planning in the urbanization of finance. As we will discuss in the next chapter, investigating these entanglements involves going beyond the prevailing research on the circuits and mechanisms through which finance capital has effectively penetrated into cities (i.e. large urban development projects and the transformation of land into a financial asset; infrastructure management and finance; and the changing relationship between cities and capital markets under urban austerity). Instead, an alternative research program requires fleshing out the variegated entanglements between finance and the multiscalar and contested generation, extraction, circulation and consumption of value that is increasingly taking place in city space itself. This is a particularly promising approach for cities and metropolitan areas in the Global South with relatively unconsolidated capital markets and institutions, which reinforces the previously mentioned case for a closer articulation between political economy-oriented strands of work and social studies of finance in the investigation of the "making of urban financialization".

Finally, although work on specific cities and sectors in Asia and Latin America has increased (particularly housing and real estate and, to a lesser extent, urban infrastructure), the research program on financialization has been criticized considering it has by and large been structured around and inspired by the urban geographies of Anglo-Saxon and North Atlantic capitalism. There has been little work on urban financialization in the Global South that goes beyond particular cities, regions or thematic areas of interest such as housing and real estate. The development of research aimed at the design of theory that goes beyond case studies, which is embedded within a perspective of the Global South, has proven challenging. This challenge can also not be dissociated from the wider infrastructure networks that guide the circulation of international knowledge, which tend to be dominated by academic and professional journals, epistemic communities, funding agencies and evaluation metrics, by and large established in the Global North (McCann, 2011; Watson, 2016).

Broadening the perspective of studying financialization embedded within the urban geography of the North requires more than analyzing contextual differences between the North and the Global South in terms of the latter's less-consolidated institutions, more intense role of the state in the rolling out of credit and capital markets and the widespread presence of governance failure (Bakker et al., 2008). Although all of this is evidently relevant, the starting point of this book will be that an investigation focussed on the specificities of the South provides insights into understanding the interdependent pattern of financialization itself at the global scale.

To clarify the argument and see what is at stake, it should be recognized that some of the scarce work that has related financialization to the Global South has done so through a conceptual lens of "subordinated-dependent" financialization (Paulani, 2011; Lapavitsas, 2013). This echoes earlier debates on the centre-periphery nexus, which occurred in the mid-20th century. The conceptual under-pinning of the centre-periphery model was structural dependency theory, which had been disseminated by intellectuals such as Prebish and Myrdal and institu-tions such as the United Nations Economic Commission for Latin America and the Caribbean (Bielschowsky, 1988). According to dependency theory, periph-eral countries were facing continuously deteriorating terms of trade in light of their export specialization in un-processed, resource-intensive export products. Without explicit strategies aimed at industrialization and import-substitution, the periphery was destined to fall behind the development trajectories of industrial-ized countries.

The literature on subordinated financialization revisits dependency theory but updates it within the perspective of a hierarchical international monetary order dominated by the US as holder of the informal world currency. The key argu-ment is that money has been "flowing uphill", particularly considering the Global South's costly accumulation of international dollar reserves and the outflow of its domestic savings in order to finance the US and other advanced market econo-mies' trade deficits (Lapavitsas, 2013; Aglietta, 2018).

As will be discussed in more detail in the next chapter, it is challenging to analyze contemporary financialization in terms of an updated centre-periphery dependency model structured around the hegemony of the dollar. First, both within the centre and the periphery, there are significant country differences in respect to their insertion into the world financial system, which suggest that money is both flowing uphill and downhill. A more fundamental problem, how-ever, is the representation of the world monetary system in terms of the notion of the periphery's dependency on the informal world money and other international key reserves held by the countries from the centre. Instead, the prevailing system of global monetary governance should be characterized in terms of its "interde-pendent" (albeit asymmetric and highly unstable) character, as illustrated by the multiple entanglements between the USA and the rest of the world in matters of trade, finance and exchange rate policies. Third, the centre-periphery metaphor is framed within the container of the national space economy. As such, it by and large fails to recognize the explicitly relational-scalar and spatial dimensions that are behind contemporary financialization, which imply a key role for cities and metropolitan areas. More specifically, the national accounting statistics regard-ing current and capital accounts and the aggregate direction of flows of funds between countries at the global level do not reveal the multiple articulations between national states and cities as strategic – albeit frequently contradictory – arenas in the mobilization of cash and credit-money, loanable funds and savings (both domestic and international) for investments. Such a perspective requires investigating how states have engaged their national space economies and cities in the production and management of recurrent crises that are a part of the prevailing international monetary order. In that sense, Brazil is an emblematic example of the state's contradictory involvement with international finance, which has culminated in a consistent inflow of international savings without triggering domestic investment and urban development. Understanding how these contradictions have unfolded in the Brazilian setting helps "worlding" the phenomenon of urban financialization within its proper global and interdependent setting (Robinson, 2006).

In the next chapter we will develop the outline of an approach that takes the previously mentioned critiques seriously in order to develop a research program that articulates political economy and social studies of finance-inspired approaches, with an emphasis on cities as privileged spaces of generation and circulation of value in the Global South. The rest of the book will then illustrate the potential of such an approach on the basis of the Brazilian trajectory.

Notes

1 Wynne Godley (1997: 4) formulated the thesis of endogenous money in the following way: "Government can no more "control" stocks of either bank money or cash than a gardener can control the direction of a hosepipe by grabbing at the water jet".
2 The Report also worked creatively with Krugman's approach aimed at bringing space back into mainstream economics through the modelling of increasing returns to scale, scale economies and imperfect competition. These efforts generated what would become known as the New Economic Geography (Krugman, 1995). For a critical debate on the limits and potentials of New Economic Geography from the disciplinary perspective of geographers see Martin and Sunley (1996).
3 The discussion of the Pirelli case in Milan, Italy (see section 1.2), provides an emblematic example of the active involvement of industrial capital itself in the mobilization and financial leverage of its land assets (Kaika and Rugierro, 2016).
4 See, in that respect, Harvey's ideas on capital switching from the primary (productive) to the secondary (built environment) circuit. More recent work on the 2007 global mortgage crisis has shown that finance capital was also instrumental in the constitution of new interdependencies between the secondary and the tertiary circuit (related to Research and Development and the reproduction of life through private plans for complementary retirement and health services, for example). The real estate bubble in the USA triggered an increasing debt level of the more vulnerable segments of the population, who re-mortgaged their houses in order to contract complementary insurance services for health care and retirement (Aalbers, 2012).
5 Such as reversed mortgages, mortgages linked to life insurance services, micro-finance and consigned credit.
6 The French Regulation School was criticized in view of its somewhat functionalist interpretation of history and capitalist dynamism (Amin, 1994). More recent regulationist-inspired work on issues such as urban neoliberalization, state rescaling and restructuring (Jessop, 2000; Brenner, 2004, 2009; and Brenner and Theodore, 2002) has allowed for contingency in the investigation of the contradictory trajectories of state rollback and rollout in the demise of Fordist welfare regimes. For a similar discussion in the Brazilian scenario see Klink (2013).
7 In operational terms this means maximizing, in present values of the moment of decision-making, the sum of projected dividends and capital gains for shareholders (i.e. difference between acquisition and sale price of shares).

8 As will be discussed in section 1.2, the regulationist arguments structured around the transition from Fordist-Keynesian to property regimes also influenced the debates on financialization in cities, as exemplified by analyses on "asset price urbanism" in Ireland (Byrne, 2016), the emerging homeownership society in Spain and Madrid (López and Rodríguez, 2011) and the role of the developmental state in triggering speculative urbanism in Seoul (Korea) (Shin and Kim, 2016), among some of the examples.

9 Moreover, a full-fledged finance-led regime has hollowed out another pillar of the Keynesian-welfare regime, that is, monetary and fiscal policies aimed at full employment.

10 According to Lavoie (2006: 54), Post-Keynesian thought on monetary theory predates Keynes himself, considering that, already in the 1830 and 1840s, classical economists such as John Fullarton and Thomas Tooke developed ideas on endogenous money.

11 The concept of credit rationing is worked out by the difference between "notional demand" for credit (from both creditworthy clients and those who are not) and the "effective demand", which only includes potential creditworthy clients (Lavoie, 2006: 70).

12 Erturk et al. (2008) also recall that the centre-left, liberal-collectivist critique on the role of the rentier-financier goes back to earlier Marxist writing by Hilferding (1910) on founder's or promoter's profit from initial public offerings and subsequent trading of paper in secondary markets, aimed at maximizing financial profits (See also Chapter 2). Its difference with the Marxian tradition, of course, was that liberal-collectivists accepted private property and wage labour but argued in favour of mechanisms of collective control aimed at stabilizing markets and guaranteeing social security.

13 Nevertheless, Erturk et al. (2008) and Froud et al. (2006) argue that the authors associated with principal-agency theory have had difficulties to empirically confirm the shareholder value dominance, for example, by showing consistently rising share prices and increasing dividends.

14 As reminded by these authors, ROCE is the ratio between returns and capital employed. Even considering that in specific sectors returns have increased, this has been accompanied by aggressive mergers and acquisition strategies coordinated by investment bankers, which effectively also raised the denominator of the ratio.

15 Nevertheless, as analyzed by Froud et al. (2006), performativity is never complete, considering that the 1987 crash generated discrepancies that were not predicted by the model. These authors also remind us, on the basis of Thrift's work (2001) on the "new economy" dot.com model, that the performativity of the Science, Technology and Knowledge economy was likewise incomplete, considering it was unable to perform on the unexpected bursts of the high-tech bubble in the early 2000s.

16 To be clear, we are paraphrasing Christopher's (2014a) dialectics structured around the notion "from Marx to markets and from markets back to Marx", considering that the other strands of political economy we have discussed earlier on (i.e. regulation theory and Post-Keynesianism) offer similar perspectives for complementarity with the concept of performativity.

17 Bayliss (2014) observes that these narratives are also driven by international funding agencies such as the World Bank and the International Finance Company.

18 In Chapter 5 we will discuss the experience of the listed mixed-capital water and sewage company of the state of São Paulo (SABESP), which has gradually adopted principles of corporatized shareholder governance.

19 In Chapter 7 we will develop a more detailed analysis on the rescaling and reframing of the capital market-austerity nexus and the emergence of financialized urban governance in Brazilian cities.

References

Aalbers MB (ed.) (2012) *Subprime cities: The political economy of mortgage markets.* Oxford: Wiley Blackwell.

Aalbers MB (2019) Financial geographies of real estate and the city: A literature review. In: *Financial geography working paper series*. ISSN 2515–0111. Leuven: University of Leuven.

Aalbers MB and Christophers B (2014) Centring housing in political economy. *Housing, Theory and Society* 31(4): 373–394.

Aglietta M (2010) Shareholder value and corporate governance: Some tricky questions. *Economy and Society* 29(1): 146–159.

Aglietta M (2018) *Money: 5,000 years of debt and power*. London and New York: Verso.

Alonso W (1964) *Location and land use*. Cambridge, MA: Harvard University Press.

Amin A (1994) *Post-Fordism: A reader*. Oxford: Blackwell.

Arrighi G (1994) *The long twentieth century: Money, power and the origins of our times*. London: Verso.

Ashton P; Doussard M and Weber R (2016) Reconstituting the state: City powers and exposures in Chicago's infrastructure leases. *Urban Studies* 53(7): 1384–1400.

Aversa M; Empinotti V and Klink J (2018) Água: mercadoria, bem comum ou direito? Algumas contradições na implementação da política pública de direitos humanos de acesso à água e ao saneamento. *Waterlat-Gobacit Network Working Papers*. Thematic Area Series-TA3 5(3): 31–52.

Bakker K (2007) The "commons" versus the "commodity": Alter-globalization, anti-privatization and the human right to water in the global South. *Antipode* 39(3): 430–455.

Bakker K (2013) Neoliberal versus postneoliberal water: Geographies of privatization and resistance. *Annals of the Association of American Geographies* 103(2): 253–260.

Bakker K; Kooy M; Shofiani NE and Martijn EJ (2008) Governance failure: Rethinking the institutional dimensions of urban water supply to poor households. *World Development* 36(10): 1891–1915.

Baran PA and Sweezy PM (1966) *Monopoly capital*. New York: Monthly Review Press.

Bayliss K (2014) The financialization of water. *Review of Radical Political Economics* 46(3): 292–307.

Berle AA and Means GC (1932) *The modern corporation and private property*. Revised Edition 1968. New York: Harcourt, Brace and World, Inc.

Berndt C and Boeckler M (2009) Geographies of circulation and exchange: Construction of markets. *Progress in Human Geography* 33(4): 535–551.

Bielschowsky R (1988) *Pensamento econômico brasileiro. O ciclo ideológico do desenvolvimentismo*. Rio de Janeiro: Contraponto.

Black F and Scholes M (1973) The pricing of options and corporate liabilities. *Journal of Political Economy* 81(3): 637–654.

Blanchard OJ (1980) The monetary mechanism in the light of rational expectations. In: Fischer F (ed.) *Rational expectations and economic policy*. Chicago: University of Chicago Press, pp. 75–116.

Block F (2016) Introduction to the special issue: The contradictory logics of financialization: Bringing together Hyman Minsky and Karl Polanyi. *Politics & Society* 44(1): 3–13.

Boag G and McDonald DA (2010) A critical review of public-public partnerships in water services. *Water Alternatives* 3(1): 1–25.

Botein H and Heidkamp CP (2013) Tax lien sales as local neoliberal governance strategy: The case of Waterbury, Connecticut. *Local Economy* 28(5): 488–498.

Boyer R (2000a) Is a finance-led growth regime a viable alternative to Fordism? A preliminary analysis. *Economy and Society* 29(1): 111–145.

Boyer R (2000b) The political in the era of globalization and finance: Focus on some Régulation School Research. *International Journal of Urban and Regional Research* 24(2): 274–322.

Boyer R (2009) *A Teoria da Regulação. Os Fundamentos.* São Paulo: Estação Liberdade.

Braudel F (1982) *Civilization and capitalism, 15th–18th century: The wheels of commerce.* Translated by Sian Reynolds. Berkeley: University of California Press.

Brenner N (2004) *New state spaces: Urban governance and the rescaling of statehood.* Oxford: Oxford University Press.

Brenner N (2009) Open questions on state rescaling. *Cambridge Journal of Regions, Economy and Society* 2: 123–139.

Brenner N and Theodore N (2002) *Spaces of neoliberalism: Urban restructuring in Western Europe and North America.* Oxford: Blackwell.

Bresnihan P (2016) The biofinancialization of Irish Water: New advances in the neoliberalization of vital services. *Utilities Policies* 40: 115–124.

Britto AL and Rezende SC (2017) A política pública para os serviços urbanos de abastecimento de água e esgotamento sanitário no Brasil: financeirização, mercantilizacão e perspectivas de resistência. *Cadernos Metrópole* 19(39): 557–581.

Brueckner JK (2001) Tax increment financing: A theoretical inquiry. *Journal of Public Economics* 81(2): 321–343.

Byrne M (2016) "Asset price urbanism" and financialization after the crisis: Ireland's National Asset Management Agency. *International Journal of Urban and Regional Research* 40(1): 31–45.

Callon M (1998) Introduction: The embeddedness of economic markets in economics. In: Callon M (ed.) *The law of the markets.* Oxford: Blackwell.

Canettieri T (2017) A produção capitalista do espaço e a gestão empresarial da política urbana. *Revista brasileira de estudos urbanos e regionais* 19(3): 513–529.

Charnock G; Purcell TF and Ribera-Fumaz R (2014) City of rents: The limits to the Barcelona model of urban competitiveness. *International Journal of Urban and Regional Research* 38(1): 198–217.

Chesnais F (ed.) (1996) *La mondialisation financière: Genèse, enjeux et coûts.* Paris: Syros.

Chiapello E (2015) Financialisation of valuation. *Hum Studies* 38: 13–35.

Christophers B (2014a) From Marx to market and back again: Performing the economy. *Antipode* 57(1): 12–20.

Christophers B (2014b) Wild dragons in the city: Urban political economy, affordable housing development and the performative world-making of economic models. *International Journal of Urban and Regional Research* 38(1): 79–97.

Christophers B (2015) From financialization to finance: For "de-financialization". *Dialogues in Human Geography* 5(2): 229–232.

Christophers B (2017) The state and financialization of public land in the United Kingdom. *Antipode* 49(1): 62–85.

Clarke DB and Bradford MG (1998) Public and private consumption and the city. *Urban Studies* 35(5/6): 865–888.

Coase RH (1960) The problem of social cost. *The Journal of Law & Economics* 3: 1–44.

Crosby N and Henneberry J (2016) Financialization, the valuation of investment property and the urban built environment in the UK. *Urban Studies* 53(7): 1424–1441.

Deruytter L and Derudder B (2019) Keeping financialization under the radar: Brussels Airport, Macquarie Bank and the Belgian politics of privatized infrastructure. *Urban Studies* 56(7): 1347–1367.

DiPasquale D and Wheaton W (1996) *Urban economics and real estate markets*. Upper Saddle River, NJ: Prentice Hall.

Eckhard H; Dodig N and Budyldina N (2014) *Financial, economic and social systems: French regulation school, social structures of accumulation and post-Keynesian approaches compared*. Working Paper No. 134. Institute for International Political Economy, Berlin.

Erturk I; Froud J; Johal S; Leaver A and Williams K (2008) *Financialization at work: Key Texts and Commentaries*. London and New York: Routledge.

Fama E (1970) Efficient capital markets: A review of theory and empirical work. *Journal of Finance* 25(2): 383–417.

Faulhaber GR and Baumol WJ (1988) Economists as innovators: Practical products of theoretical research. *Journal of Economic Literature* 26(2): 577–600.

Fields D (2014) Contesting the financialization of urban space: Community organizations and the struggle to preserve affordable rental housing in New York City. *Journal of Urban Affairs* 37(2): 144–165.

Fix M (2007) *São Paulo Cidade Global: Fundamentos financeiros de uma miragem*. São Paulo: Boitempo.

Friedman M (1968) The role of monetary policy. *American Economic Review* 58(1): 1–17.

Froud J; Haslam C; Johal S and Williams K (2001) Financialization and the coupon pool. *Gestão & Produção* 8(3): 271–288.

Froud J; Johal S; Leaver A and Williams K (2006) *Financialization and strategy: Narrative and numbers*. London and New York: Routledge.

Godley W (1997) *Macroeconomics without equilibrium or disequilibrium*. Working Paper No. 205. The Levy Economics Institute, Annadale-on-Hudson, NY.

Gotham K (2009) Creating liquidity out of spatial fixity: The secondary circuit of capital and the subprime mortgage crisis. *International Journal of Urban and Regional Research* 33(2): 355–371.

Graham S and Marvin S (2001) *Splintering urbanism: Networked infrastructures, technological mobilities and the urban condition*. London and New York: Routledge.

Granovetter M (1985) Economic action and social structure: The problem of embeddedness. *American Journal of Sociology* 91: 481–510.

Guironnet A; Attuyer K and Halbert L (2016) Building cities on financial assets: The financialization of property markets and its implications for city governments in the Paris city region. *Urban Studies* 53(7): 1442–1464.

Guironnet A and Halbert L (2015) *Urban development projects, financial markets, and investors: A research note*. Chairville: École des Ponts Paritech.

Guy S; Henneberry J and Rowley S (2002) Development cultures and urban regeneration. *Urban Studies* 39(7): 1181–1196.

Haila A (1988) Land as a financial asset: The theory of urban rent as a mirror of economic transformation. *Antipode* 20(2): 79–100.

Haila A (2016) *Urban land rent: Singapore as a property state*. West Sussex: Wiley-Blackwell.

Halbert L and Attuyer K (2016) Introduction: The financialisation of urban production: Conditions, mediations and transformations. *Urban Studies* 53(7): 1347–1361.

Hicks J (1974) *The crisis in Keynesian Economics*. Oxford: Basil Blackwell.

Hildyard N (2016) *Licensed larceny*. Manchester: Manchester University Press.

Hilferding R (1910) *Finance capital*. London: Routledge & Kegan Paul (1981 Edition).

Hall S (2010) Geographies of money and finance I: Cultural economy, politics and place. *Progress in Human Geography* 35(2): 234–245.

Harcourt GC (2006) *The structure of post-Keynesian economics: The core contributions of the pioneers.* Cambridge: Cambridge University Press.

Harvey D (1989) From managerialism to entrepreneurialsm: The transformation in urban governance in late capitalism. *Geografiska Annaler: Series B, Human Geography* 71(1): 3–17.

Harvey D (2009) Assessments: Reshaping economic geography: The world development report 2009. *Development and Change* 40(6): 1269–1277.

Harvey D (2013) *Os Limites do Capital.* Translated by Magda Lopes. São Paulo: Boitempo.

Helleiner E (1994) *States and the re-emergence of global finance: From Bretton Woods to the 1990s.* Ithaca and London: Cornell University Press.

Henneberry J and Roberts C (2008) Calculated inequality? Portfolio benchmarking and regional office property investment in the UK. *Urban Studies* 45(5&6): 1217–1241.

Henrikson LF (2009) *Are financial markets embedded in economics rather than society? A critical review of the performativity thesis.* DIIS Working Paper. Danish Institute for International Studies, Copenhagen.

Jensen M and Meckling W (1976) Theory of the firm: Managerial behaviour, agency costs and ownership structure. *Journal of Financial Economics* 3(4): 305–360.

Jessop B (2000) The crisis of the national spatio-temporal fix and the tendential ecological dominance of globalizing capital. *International Journal of Urban and Regional Research* 24(2): 323–360.

Johnson B (2016) Atlantic City takeover plan moves forward despite uproar. *NJ.com*, 10 March. Available at: https://www.nj.com/politics/2016/03/atlantic_city_takeover_plan_moves_forward_despite.html (accessed 2 December 2019).

Kaika M and Rugierro L (2016) Land financialization as a "lived" process: The transformation of Milan's Bicocca by Pirelli. *European Urban and Regional Studies* 23(1): 3–22.

Kaldor N (1985) *Economics without equilibrium.* Armonk, NY: ME Sharpe.

Kalecki M (1971) *Selected essays on the dynamics of the capitalist economy.* Cambridge: Cambridge University Press.

Keynes JM (1930) *A treatise on money.* 2 vols. London: Macmillan.

Keynes JM (1937) The "ex-ante" theory of the rate of interest. *Economic Journal* 47: 663–669.

Kirkpatrick LO (2016) The new urban fiscal crisis: Finance, democracy, and municipal debt. *Politics & Society* 44(1): 45–80.

Klink J (2013) The hollowing out of Brazilian metropolitan governance as we know it: Restructuring and rescaling the developmental state in metropolitan space. *Antipode* 46(3): 629–649.

Klink J and Denaldi R (2014) On financialization and state spatial fixes in Brazil: A geographical and historical interpretation of the housing program My House My Life. *Habitat International* 44: 220–226.

Klink J; Empinotti V and Aversa M (2019) On contested water governance and the making of urban financialization: Exploring the case of metropolitan São Paulo, Brazil. *Urban Studies*. Online First. Forthcoming.

Klink J and Stroher L (2017) The making of urban financialization? An exploration of Brazilian urban partnership operations with building certificates. *Land Use Policy* 69: 519–528.

Krugman P (1995) *Development, geography and economic theory.* Cambridge, MA: The MIT Press.

Langley P (2004) In the eye of "the perfect storm": The final salary pensions crisis and the financialisation of anglo-American capitalism. *New Political Economy* 9(4): 539–538.

Lapavitsas C (2013) *Profiting without producing: How finance exploits all of us*. London and New York: Verso.

Latour B (2005) *Reassembling the social: An introduction to actor-network theory*. Oxford: Oxford University Press.

Lavoie M (2006) *Introduction to Post-Keynesian economics*. New York: Palgrave MacMillan.

Lazonick W and O'Sullivan M (2000) Maximizing shareholder value: A new ideology for corporate governance. *Economy and Society* 29(1): 13–35.

Lipietz A (1987) *Mirages and miracles: Crisis in global Fordism*. London: Verso.

López I and Rodríguez E (2011) The Spanish model. *New Left Review* 69: 5–29.

MacKenzie D (2005) *An engine, not a camera: How financial models shape markets*. Cambridge, MA: The MIT Press.

March H and Purcell T (2014) The muddy waters of financialization and new accumulation strategies in the global water industry: The case of AGBAR. *Geoforum* 53: 11–20.

Martin R (2002) *The financialization of daily life*. Philadelphia: Temple University Press.

Martin R and Sunley P (1996) Paul Krugman's geographical economics and its implications for regional development theory: A critical assessment. *Economic Geography* 72(3): 259–292.

McCann E (2011) Urban policy mobilities and global circuits of knowledge: Toward a research agenda. *Annals of the Association of American Geographers* 101(1): 107–130.

Minsky HP (1981) *Can "it" happen again? Essays on instability and finance*. Armonk, NY: ME Sharpe.

Mosciaro M and Pereira A (2019) Reinforcing uneven development: The financialization of urban redevelopment projects. *Urban Studies*. Online First. Available at: https://doi.org/10.1177/004209801982942 (Accessed 1 April 2019).

Moulaert F (2005) Institutional economics and planning theory: A partnership between ostriches? *Planning Theory* 4(1): 21–32.

Muth R (1969) *Cities and housing*. Chicago: University of Chicago Press.

Newborne P and Mason N (2012) The private sector's contribution to water management: Re-examining corporate purposes and company roles. *Water Alternatives* 5(3): 603–618.

Nitzan J and Bichler S (2009) *Capital as power: A study of order and creorder*. London and New York: Routledge.

North D (1990) *Institutions, institutional change and economic performance*. Cambridge: Cambridge University Press.

O'Neill P (2018) The financialisation of urban infrastructure: A framework of analysis. *Urban Studies*. Online First. Available at: https://doi.org/10.1177%2F0042098017751983 (Accessed 21 September 2018).

Pacewicz J (2013) Tax increment financing, economic development professionals and the financialization of urban politics. *Socio-Economic Review* 11: 413–440.

Pacewicz J (2016) The city as a fiscal derivative: Financialization, urban development, and the politics of earmarking. *City & Community* 15(3): 264–288.

Paulani LM (1991) *Do conceito de dinheiro e do dinheiro como conceito*. Tese de Doutorado. Faculdade de Economia, Administração e Contabilidade da Universidade de São Paulo, São Paulo.

Paulani LM (2011) A inserção da economia brasileira no cenário mundial: uma reflexão sobre o papel do Estado e sobre a situação atual real à luz da história. In: *Logros e Retos del Brasil Contemporâneo*. Cidade de México, México, 24 a 26 de Agosto de 2011. UNAM.

Peck J (2017a) Transatlantic city, part 1: Conjunctural urbanism. *Urban Studies* 54(1): 4–30.

Peck J (2017b) Transatlantic city, part 2: Late entrepreneurialism. *Urban Studies* 54(2): 327–363.

Peck J and Whiteside H (2016) Financializing Detroit. *Economic Geography* 92(3): 235–268.

Polanyi K (1997) *The great transformation: The political and economic origins of our time*. Boston, MA: Beacon Press.

Pryke M and Allen J (2017) Financialising urban water infrastructure: Extracting local value, distributing value globally. *Urban Studies*. Online First: 1–21. Available at: https://doi.org/10.1177/0042098017742288 (Accessed 2 April 2019).

Robinson J (2006) *Ordinary cities: Between modernity and development*. London: Routledge.

Rolnik R (2018) *Urban warfare: Housing under the empire of finance*. London: Verso.

Rooyen vC and Hall D (2007) *Public is as private Does: The confused case of Rand Water in South Africa*. Municipal Services Project. Occasional Paper Series No. 15. Logo Printers, Cape Town.

Rouanet H and Halbert L (2016) Leveraging finance capital: Urban change and self-empowerment of real estate developers in India. *Urban Studies* 53(7): 1401–1423.

Rutland T (2010) The financialization of urban redevelopment. *Geography Compass* 4(8): 1167–1178.

Sandroni P (2010) A new financial instrument of value capture in São Paulo: Certificates of additional construction potential. In: Ingram GK and Hong YH (eds.) *Municipal revenues and land policies*. Cambridge, MA: Lincoln Institute of Land Policy, pp. 218–236.

Sanfelici D (2013) Financeirização e a produção do espaço urbano no Brasil: uma Contribuição ao debate. *Eure* 39(118): 27–46.

Savini F and Aalbers M (2016) The de-contextualisation of land use planning through financialisation: Urban redevelopment in Milan. *European Urban and Regional Studies* 23(4): 878–894.

Shin HB and Kim SH (2016) The developmental state, speculative urbanisation and the politics of displacement in gentrifying Seoul. *Urban Studies* 53(3): 540–559.

Steindl J (1952) *Maturity and stagnation in American capitalism*. Oxford: Basil Blackwell.

Swyngedouw E; Moulaert F and Rodriguez A (2002) Neoliberalization in Europe: Large scale urban development projects and the new urban policy. *Antipode* 34(3): 542–577.

Tawney RH (1923) *The acquisitive society*. London: Bell and Sons.

Theurillat T and Vera-Büchel N (2016) Commentary: From capital lending to urban anchoring: The negotiated city. *Urban Studies* 53(7): 1509–1518.

Thrift N (2000) Pandora's box? Cultural geographies of economies. In: Clark G; Feldman M and Gertler M (eds.) *The Oxford handbook of economic geography*. Oxford: Oxford University Press, pp. 689–704.

Thrift N (2001) "It's the romance not the finance that makes the business worth pursuing": Disclosing a new market culture. *Economy and Society* 30(4): 412–432.

Torrance MI (2008) Forging glocal governance? Urban infrastructures as networked financial products. *International Journal of Urban and Regional Research* 32(1): 1–21.

Ülgen F (2014) Financialized capitalism and the irrelevance of self-regulation: A Minskyian analysis of systemic viability. In: *12th International Post Keynesian Conference*, Kansas City, Missouri, USA, September 25–28.

Van Loon J and Aalbers MB (2017) How real estate became "just another asset class": The financialization of the investment strategies of Dutch institutional investors. *European Planning Studies* 25(2): 221–240.

Van Loon J; Oosterlynck S and Aalbers MB (2018) Governing urban development in the low countries: From managerialism to entrepreneurialism and financialization. *European Urban and Regional Studies*. Online First. Available at: https://doi.org/10.1177/0969776418798673 (Accessed 1 April 2019).

Watson V (2016) Shifting approaches to planning theory: Global North and South. *Urban Planning* 1(4): 32–41.

Weber (2010) Selling city futures: The financialization of urban redevelopment policy. *Economic Geography* 86(3): 251–274.

Weber (2015) *From boom to bubble: How finance built the new Chicago*. Chicago and London: The University of Chicago Press.

Whiteside H (2019) Advanced perspectives on financialised urban infrastructures. *Urban Studies*. Online First. Available at: https://doi.org/10.1177/0042098019826022 (Accessed 3 April 2019).

Wijburg G (2018) Privatised Keynesianism and the state-enhanced diversification of credit: The case of the French housing market. *International Journal of Housing Policy*. Online First. Available at: www.tandfonline.com/doi/full/10.1080/19491247.2017.1397 926 (Accessed 5 April 2019).

Williamson O (1985) *The economic institutions of capitalism*. New York: The Free Press.

World Bank (2009) *World development report 2009: Reshaping economic geography*. Washington, DC: The World Bank.

van der Zwan N (2014) Making sense of financialization. *Socio-Economic Review* 12: 99–129.

2 State spaces, financialization of cities and the urbanization of finance in the Global South

2.1 State rescaling, space and the financialization of cities

2.1.1 Introduction

Cities are more than dense land and infrastructure networks, pools of labour and agglomerations of economic activity and creativity. The design of new theoretical approaches on financialization requires acknowledging cities' complexity as privileged spaces for the generation, extraction/appropriation and circulation of value and (collective) consumption and, ultimately, for the (re)production of daily life.

As a matter of fact, this recognition was the entrance point for much of the Marxian critical urban studies from the 1970s onwards, undertaken by authors such as Harvey (1973, 1982), Soja (1989), Castells (1977) and Lefebvre (1974), among some of the well-known examples. Their work placed cities at the centre-stage of an investigation of the contradictory spatial trajectory of capitalism. Despite the internal differences in approaches, critical urban studies argued that, in addition to the "historical" evolution of class conflict, "space" was crucial in understanding the dynamics of capitalist accumulation and reproduction itself.

More recently, authors such as Brenner (2004) and Brenner et al. (2010) have extended this research program by creatively establishing bridges between Lefebvre's work on the right to the city, both "as a complaint and a demand for something better" (Harvey, 2014: 13) and political economy-inspired investigations of neoliberal governance, state spatial restructuring and rescaling. To be specific, Brenner's analysis was structured around the transformation of what he called spatial Keynesianism – characterized by hierarchical, nationally scaled and homogeneous state strategies aimed at redistribution within domestic economies – into state spatial regimes targeted at the competitive insertion of cities and metropolitan areas into the global economy.

In a way, Brenner's approach built on Harvey's earlier description of the emergence of urban entrepreneurialism and was grounded in an analysis on the rescaling and restructuring of state spatial projects (i.e. the territorial organization of the state, through (de)centralized or homogenous/customized institutional arrangements) and state spatial strategies (i.e. related to how the state effectively shapes its territorial interventions through regulation and investment) in the social constitution of variegated patterns of urban neoliberalization (Brenner et al., 2010). Likewise, this work responded to criticism that was raised about the first wave of regulationist political economics regarding its somewhat linear interpretation of history.[1] Thus, neoliberalization should not be interpreted in terms of a straightforward process of penetration of the private sector into the planning and management of state assets. Instead, the neoliberalization project was advancing into our daily lives through successive and contradictory crises-driven rounds of rollback and rollout of the state, which gradually succeeded in consolidating its hegemony (Brenner and Theodore, 2002).

This literature also established conceptual linkages with areas such as critical geography, which contributed to both flesh out the significance and nature of scaling and rescaling under contemporary capitalism and its implications for a better understanding of the neoliberalization of urban spaces. This articulation with critical geography consolidated a relational perspective on scale, according to which the latter should be analyzed in terms of an arena that is contested by state and non-state actors alike, which all strive to fill it in according to their particular projects and interests (Swyngedouw, 1997; Brown and Purcell, 2004; Klink, 2014). As such, there was nothing inherent to particular (global, national, regional or local) scales; instead, its understanding required a more detailed investigation of the projects and strategies of social actors who tried to hollow out, or fill in, specific scales according to their projects. Interestingly enough, this conflict-driven, relational reading of scaling and rescaling also provided urban political economy with more contingency and sensitivity to place- and context-specific experiences. With this more open-ended and relational understanding, Brenner conceptualized the urban question as a scalar question (Brenner, 2000).

Likewise, Peck (2018) argued that a variegated and contextually situated approach to state, space and scale would also strengthen the complementarities between urban political economy and other perspectives within critical urban studies. In relation to the latter, for example, cities were not only recognized as being the privileged locus of class conflict but also concentrated important dynamics of racial, gender and ethnic segregation as part of the "urban problematic" (Dymski, 2009: 427; Massey, 2007). Moreover, influential work was being undertaken along the lines of post-structural (Barnes, 1996)[2] and post-colonial studies. These approaches emphasized the complexity and specificity of cities, which challenged "pre-fixed" narratives, particularly those originating from the North. According to Peck, the post-structural and post-colonial literature had successfully questioned the capacity of general conceptual frameworks "from the North" to provide comparability and commensurability among cities, particularly those in the Global South. Nevertheless, while this critique was welcome, it had also generated side effects. One of them was a shift within critical studies towards an emphasis on the unique character of the urban experience in general, and that from the Global Urban South in particular. Perhaps as an unintended consequence, the area of critical urban studies was now at risk of hollowing out its capacity, or perhaps even its ambition, to look beyond specific cases in order to discern broader patterns (Robinson, 2011a, 2011b). Thus, a more open-ended, variegated approach towards political economy would help to establish bridges within critical urban studies with an open eye for the possible linkages between context-specific city trajectories and "revisable claims of theory" (Peck, 2017a: 4).

In spite of its considerable achievement in building a better understanding of the role of cities in times of globalization and neoliberalization, as well as of the diversity and capacity of cities to "invent their own praxis",[3] it should nevertheless be recognized that critical urban studies, with a few notable exceptions, has been rather silent on issues related to money, credit and finance.[4] This silence should be seen in light of its priority on the investigation of globalization, neoliberalization of state spaces and rescaling (i.e. in terms of regulation theory's lenses, the "mode of regulation"), while it implicitly put less emphasis on the analysis of economic restructuring itself (that is, into the direction of a finance-led development regime). In a way, then, critical urban studies need better links between the work it has consolidated on state spaces and the variegated patterns of neoliberalization and the emerging research agenda on financialization, which we discussed in the previous chapter.

The articulation between state spatial theory, scale and "heterodox economics' perspective on money, credit and finance" provides the first of such links, which will be discussed in two steps in this section ("State spaces, scale and money" and "States, the public fund and valuing risk in the making of financial markets" in section 2.1.2 and 2.1.3, respectively).[5] In section 2.2 we analyze the political economy of valuation as a complementary articulation, which is influenced by social studies of finance.

2.1.2 State spaces, scale and money

Our definition of "money" will follow the earlier-mentioned heterodox strands of economic thinking that were discussed in Chapter 1. Thus, as argued by Jessop (2015: 21), "money is not a thing but a fetishized social relation with the potential to generate economic, financial and fiscal crises". For the specific purposes of our discussion, it is important to distinguish between fiduciary, regulated or credit money, on the one hand, and commodity money such as bullion and gold on the other.[6] Different from commodity money and bullion, fiat and credit money do not "incorporate value" (in the Marxian sense), considering these do not embody different labour value times; "nonetheless, they do *represent or reflect value* in the sense that they comprise claims on social wealth insofar as this takes the form of an immense accumulation of commodities" (Marx, 1967, cited by Jessop, 2015: 28). The dissemination of a variety of derivatives and related financial engineering ("fictive money" in Jessop's terminology) in the post-Bretton Woods era has only increased the significance of this distinction considering the built-in capacity to accumulate contradictions and tensions within the global pyramid of money.

In a way, the monetary history has been accompanied by recurrent stages whereby the links between fiduciary and credit money and commodity money and hard cash have been occasionally loosened in order to guarantee a more elastic growth of production, commerce and trade, or to finance large-scale investments or expenses on religious or territorial wars, among some of the more frequent reasons. The bottom line here is, of course, that fiduciary and credit money have always required varying degrees of social belonging and trust, as well as conventions and institutions, which have invariably been regulated and mediated by a sovereign state. In a way, the strength of fiduciary and regulated fiat money has been umbilically linked to the legitimacy and capacity of sovereign states.

It should be stressed, however, as is also discussed by Aglietta (2018), that the variegated constitution of the state-money nexus has by no means always been a monopoly of the nation-state. From the ancient Greeks, the Roman empire to the Dutch-Italian commercial hegemony during the renaissance, cities and city-states have been privileged arenas for the design and implementation of monetary systems. This has included the provision of guarantees in order to maintain stability between a growing circulation of fiduciary and credit money in relation to the underlying base of commodity money (gold, silver, bullion and so on). The key point here is that there has always been a close relationship between the transformation of the territorial-scalar basis of the state itself and the historical evolution of monetary systems. Until the large-scale emergence of industrialization, city-states have been at the forefront in the organization of money and credit within national and international systems of trade and commerce. It was only in a subsequent stage that nation-states and modern central banking systems have gradually filled in what would become the international monetary order of the gold standard in the nineteenth, as well as its twentieth-century successor arrangements of Bretton Woods (gold-dollar standard) and, eventually, the dollar standard.

Modern twentieth-century monetary systems evolved into hierarchical arrangements, embedded within national systems of central banks as "lenders of last resort". The latter have alleviated tensions within prevailing interstate political arrangements regarding the organization of domestic money, national currencies and world money. The governance of the Bretton Woods system, which prevailed during the golden years of international capitalism between 1946–1971, was a paradigmatic example (Helleiner, 1994). It represented a political pact among members which was grounded within a collective perception that it was crucial to avoid the trauma of competitive devaluations and trade wars that marked the inter-bellum. In other words, the growth of international trade and commerce would require barriers on the international circulation of finance capital. This political pact was operationalized through strict regulation of national capital markets, combined with a system of fixed exchange rates, which was anchored on the convertibility (at a pre-established rate of exchange) between the emerging informal world currency (the US dollar) and the gold reserves. In a way, Bretton Woods represented a second-best governance solution which only partially compensated for the lack of a supranational monetary authority and an effective world currency (as opposed to an informal one). The ideal solution, as envisaged by Lord Keynes during the negotiations in Bretton Woods, would have provided international emergency liquidity and a "supra-national" global lender of last resort in order to "symmetrically" divide the burden of adjustment between debtor and creditor countries so as to avoid insufficient global effective demand.

Instead, under pressure from the US, Bretton Woods eventually adopted an "intergovernmental" system anchored to the dollar and the international gold reserves. It placed an asymmetric burden of adjustment on debtor countries, which would generate a built-in deflationary tendency and insufficient global effective demand.

While Bretton Woods represented an essentially political pact among its members aimed at restricting exchange rate volatility and international mobility of capital, its hollowing out in the 1970s should therefore not be interpreted as a "metaphysical" landing of global finance into the prevailing monetary order. Instead, the deregulation of national capital markets, liberalization of international capital flows and the demise of the system of fixed exchange rates, including the convertibility between the dollar and the gold reserves, represented a meticulously filling in and rolling out of a finance-driven regime, which was mediated by national states in general, and the US and the UK in particular (Helleiner, 1994). The resulting tensions and contradictions have nevertheless unfolded on multiples scales.

For the purpose of our argument, the demise of Bretton Woods has aggravated and exposed the mismatches between states, scales and space in the planning and management of the global monetary order. As exemplified by the global financial subprime crisis, this has been accompanied by a gradual re-emergence of the city-regional and metropolitan scale in the design, proliferation as well as absorption of a variety of forms of fiduciary-fictitious and credit money. The planning and finance of big cities and metropolitan areas have represented important arenas for the creation of innovative financial engineering and dissemination of credit and fictitious money. Moreover, as will be discussed in more detail in section 2.3, within national territories cities also represent spaces for the absorption of international savings (that is, foreign investors buying domestic assets) and domestic wealth flowing in from smaller and economically less dynamic areas of the country (which are net buyers of goods and services from metropolitan areas). This intense creation and circulation of assets in urban and metropolitan spaces has set the stage for loosening the connections with hard cash and the high-powered monetary base, and, as such, has enabled increasingly speculative investments, particularly in institutional environments marked by light-touch regulation and loose supervision. Recalling the Post-Keynesian perspective on endogenous money, in such optimistic scenarios, the existing flow of savings doesn't represent a bottleneck on investments and economic growth. Rather, the expectations that the projected rate of return on urban investments will exceed the cost of borrowing will determine the speed and facility through which various forms of credit and fictitious money will be created and allocated into cities by actors searching for financial profits.[7]

In times of crisis, however, the connections between credit, fiduciary money and hard cash re-emerge as a growing number of contradictions and speculative booms unfold, which, paraphrasing Aglietta (2018: 1), require finance and capital markets to be saved by money. This periodically mobilizes states at the scale of national, and ultimately world markets, through periodical crisis management, particularly in light of insufficient bullion and commodity money, the lack of a world monetary authority and effectively working mechanisms to provide a lender of last resort at the global scale. Consequently, recurrent state scalar and spatial tensions – from the global to the national and local scales and upward again- are likely to occur in what has become an increasingly interdependent financial system. The national-state mediated rescue operations that emerged after the global financial meltdown have, by and large, transformed the subprime crisis into a widely disseminated regime of fiscal crisis, accompanied by calls for fiscal austerity and state-roll back. This subsequent rescaling of crises and austerity has increasingly targeted cities (e.g. EUA) and provinces (e.g. at the regional level of the autonomous communities in Spain). Thus, the state-mediated up and downscaling of the management of money, credit and crisis has become a key dimension of the post-Bretton Woods order.[8] This requires a better theoretical understanding of the links between the state and the making of credit markets, which will be the theme of the next section.

2.1.3 States, the public fund and valuing risk in the making of financial markets

A second and related dimension of heterodox economics that is relevant for the purpose of our discussion is related to the entanglements between the state, the public fund and the valuation of risk in the constitution of markets for credit and finance.

As will be discussed in more detail in section 2.2 when dealing with the capital asset pricing model, private investors adopt specific calculative practices in order to deal with risk and uncertainty in credit and capital markets. More specifically, they price and evaluate the specific risks of holding particular assets as compared to the returns that could be obtained from a hypothetical average portfolio that has been assembled in a consolidated market. However, in thin capital markets there is no anchor to make these calculations. The lack of deep and liquid capital markets implies the existence of systemic risks, which private sector actors are unlikely to absorb considering the absence of parameters or metrics to price the trade-off between risk and return.[9] At the same time, states have always performed a strategic role in the design of the social and political contract in which efficiently working arrangements for money, credit and finance are embedded. Consequently, this has required their continuous involvement in the management of systemic risk, that is, when well-functioning systems are not in place (yet), or when prevailing mechanisms threaten to collapse in light of a crisis, lack of confidence or contagion (Aglietta, 2018). As a matter of fact, the conceptual debate regarding the sharing of risks (between the state, finance capital and other intermediaries, workers and so on) and the generation and circulation of value has been surprisingly inconclusive (Christophers, 2018; Purcell et al., 2019).

In a recent paper, for example, Christophers (2018) revisits what he calls the cumbersome relation between value theory and finance from a classical Marxian perspective. According to the latter, finance is not productive and, as such, doesn't generate value, as opposed to extracting part of it within the sphere of circulation after it has been created elsewhere. Christophers argues that this perspective is problematic considering finance capital's significant role in the contemporary global economy. His way out of this impasse is to recognize the "relational" role of the valuation and repackaging of risk as a productive (i.e. labour embedding) activity per-se, and to reconcile value and finance by what he labels as "risking value theory" (Christophers, 2018: 332).

Our approach to the issue is somewhat different, however. The entrance point is the implicit role of the public fund under the earlier discussed spatial Keynesianism (Oliveira, 1988) and how it has been the object of continuous disputes in times of neoliberalization and financialization. Under spatial Keynesianism, the public fund represented a precondition for the reproduction of both capital and labour. In relation to capital, for instance, this was reflected in the picking of national champions through targeted industrial policies and the proliferation of tax incentives, among others, with impacts on the sphere of "production and circulation" of value. In relation to labour, this was evident in the mobilization of the state around its contributions to social wages through the subsidization of housing and urban infrastructure, education and health, among others, thereby directly influencing the domain of value "distribution and consumption". As highlighted by Oliveira (1988), a key element here was the long-term predictability and transparency of the Keynesian finance standard. This was labelled by him as something of a "revolution à la Copernicus", considering that under the welfare regime "the public fund is now an *ex-ante* dimension of the reproduction of each specific capitalist and the general living conditions, instead of its typical *ex-post* character under competitive capitalism" (Oliveira, 1988: 19). Moreover, somewhat different from the emphasis of Keynesian political economics on conjuncture, the public fund is a "structural element" as soon as it is recognized that the reproduction of the system depends on a variable that denies its essence: "to summarize, the public fund is *anti-value*, not in the sense that the system doesn't generate value anymore, but related to the fact that the underlying assumptions behind the reproduction of value contain, intrinsically, the more fundamental elements of the system's denial" (Oliveira, 1988: 19). Finally, the public fund under spatial Keynesianism is intrinsically linked with predictable, transparent and long-term mechanisms through which the state absorbs the risks associated with the reproduction cost of labour.

The hollowing out of spatial Keynesianism has directly affected these multiple functions of the public fund. While the various subsidies that provided an indirect salary to workers have been downsized, the state has simultaneously been involved in the gradual rolling out of institutional incentives in order to enable access of households and workers to credit. As we have seen in the previous chapter, this requires mobilizing workers around the micromanagement of risks that had previously been internalized, *ex-ante*, by the state itself. Nevertheless, there is nothing particularly obvious in the social constitution and legitimation of credit markets for specific packages of goods and services – such as housing and urban infrastructure – which have traditionally fallen under the responsibility of the state. In a way, then, this demands a different sort of risk absorption by the public fund in the making and consolidation of credit for "cumbersome commodities" such as housing, land and urban infrastructure. Updating Oliveira's (1988) analysis, in times of neoliberalization we argue that the public fund increasingly performs a key role in the constitution of a set of metrics and institutional-financial mechanisms that facilitate both the absorption of private lender's credit risk for "un-collaborative commodities" associated with urban space, as well as the gradual financialization of daily life itself. In other words, the re-emergence of global finance cannot be separated from the gradual transformation of the public fund, as anti-value under spatial Keynesianism, to fictitious value (as driven by the search for financial profits) under a finance-driven system of accumulation.

Thus, somewhat different from Christopher's argument, we argue that the systemic risk of creating and preserving good governance in credit and capital markets, particularly so in the Global South, is largely absorbed by the state and, as such, represents a qualitative transformation of the public fund from anti-value under spatial Keynesianism to state-mediated fictitious value in times of financialization. Rather than "risking value theory", what seems to be at stake here is an active involvement of the state and the public fund in the "valuating (and pricing) of risk". Under these circumstances, the public fund becomes essential to provide "value for money" to investors in the sense of disseminating an environment of predictable, low-risk and contracted income streams in order to allow the entrance of finance capital in thematic areas that previously were under the domain of developmental or welfare states (Royer, 2014).

In the case studies that will be presented in this book,[10] cities will feature prominently considering they represent key arenas and privileged spaces where the hollowing out of the public fund in its traditional role, as well as the gradual financialization of everyday life as an asset to be micro-managed by each one of us, has been unfolding more rapidly. Nevertheless, this still leaves us with two clear open questions regarding the role of cities in times of financialization.

The first one has a conceptual dimension. Considering the indivisibilities, rigidities, long payback periods and high-risk profile of urban space, how exactly are cities being reshaped and reassembled into financial assets and investment objects? What is the role of urban planning and the public fund in mediating these transformations? And what kind of potential resistances to these processes might emerge? In the next section we will approach some of these questions on the basis of a social studies of finance-inspired approach on the politics of valuation. It provides us with an additional link between critical urban studies and the literature on financialization.

The second question is related to the geography and history of the Brazilian development trajectory in general, and the entanglements between the state, money and the finance of its cities in particular, which require mediations between some of the broader theoretical claims laid out in this chapter and the specific conditions that have accompanied the (re)production of space in the country. This will be the object of analysis in Chapter 3.

2.2 The urbanization of finance

Our hypothesis here is that in countries of the Global South, planners perform particularly important relational roles in articulating the diametrically opposing metrics, conventions and representations of space that prevail in the world of finance, on the one hand, and the indivisibilities, use values and the reproduction of life that are an essential part of cities, on the other. This is what we will call the urbanization of finance. In the setting of consolidated capital markets in developed countries, this relational role is by and large performed and facilitated by financial intermediaries and institutional investors, as described by authors such as Torrance (2009) and Pryke and Allen (2017). In countries of the Global South, however, marked by relatively thin and illiquid capital markets and risk-averse investors, the capacity of finance capital to articulate diverging metrics and representations of space and deal with systemic risks is restrained and requires a more active involvement of planners and the public fund itself in order to facilitate the transformation of cities into homogenous and commensurable asset classes. However, analyzing the urban question from a scalar perspective, there is nothing particular inherent to this transformation, which is contested by actors who strive to fill in other projects for the reproduction of daily life and differential space. Therefore, a theoretical perspective on the relational role of planners enables us to investigate variegated patterns of financialization, whereby cities invent, fill in and occasionally face contestation of specific trajectories, as opposed to a global landing and anchoring of finance in local territories.

This becomes relevant when considering that there is nothing particularly inherent to the "landing and anchoring" (Theurillat and Vera-Büchel, 2016) of finance capital in cities. The transformation of cities into investment objects involves the careful mobilization of the state, at multiples scales, both in the contested social demarcation and creation of assets, as well as in the valuation and capitalization of associated income streams such as rents, which can subsequently be used in the procurement of financial profits. In what follows, we will, step-by-step, flesh out the specific interdependent "space-time" components of a political economy of valuation and pricing that is mobilized in the making of urban financialization. By doing so, we will also establish conceptual bridges between social studies of finance and political economy-inspired "midlevel concepts" (Peck, 2017a: 4), which provide a cross-fertilization between the case studies and "revisable theory claims" that are part of the argument of this book.

A first step is related to the transformation of space itself, through what is called by authors such as Birch (2017) as the process of "assetization". Assets can be bought and sold in markets (and, as such, hypothetically provide capital gains) and generate a periodical income for their owners, which is associated with the latter's exclusive right to charge potentially interested users for accessing it.[11] Assetization inevitably requires the mobilization of legal frameworks and state projects aimed at the rolling out of customized institutional mechanisms that legitimize selective ownership. The latter enables the constitution of individual property rights and the possibility to charge consumers periodic payments that are related to user frequency and intensity, providing asset owners with a contracted and predictable stream of revenues. From the perspective of investors, the total internal rate of return on assets is thus derived through a combination of "capital gains" (the difference between buying and selling price in secondary markets) and the capitalized "periodic revenue stream" that is extracted from users.

Assetization of cities requires a continuous mobilization of the state, at multiple scales, in the transformation of space and common property resources (such as nature, land, infrastructure networks and pools of labour) into explicit investment objects that generate periodic income streams. Purcell et al. (2019) take this argument one step further and relate the city-anchored revenue streams with the concepts of monopoly and differential rents from classical value theory (Harvey, 1982). According to them, spatial common pool resources such as water and nature have no value from the perspective of classical theory (considering they have not been produced and, as such, do not embody a specific amount of labour time). Consequently, while their intrinsic use value ("as gifts from god") is being transformed in market prices as revealed preferences (producing the appearance of value), a rent is actually being extracted from other places. Their description of water pricing in the generation of monopoly rents in the experience of the Thames Water Company provides a specific example of how the argument unfolds in the historical and geographic context of infrastructure finance in England.[12]

Urban assets cannot be taken for granted; in cities, as complex entanglements of common pool resources, assets do not emerge as a spontaneous result from individual rational choice aimed at the reduction of transaction costs or the resolution of the tragedy of the commons, as envisaged by strands of new institutional economics (Moulaert, 2005; North, 1990). Instead, as we will discuss in subsequent chapters, urban and environmental assetization unfolds through contested social processes whereby private and public actors alike strive to fill in the specific legal-institutional, financial and organizational parameters that guide user-access, the elaboration of contracted revenue streams and the extraction of rents. Moreover, as will be discussed in the next chapter, the historical and geographical development trajectory since colonial times has transformed Brazilian cities and metropolitan areas into privileged spaces for the generation of rents and urban founder's profits.

A second step in a political economy of pricing and valuation is related with time and risk. Any investment implies an intrinsic tension between the opportunity costs of losing liquidity in the present and the expectation of receiving additional money in the future. Normal economic science and finance theory have come up with a specific device to evaluate the time dimension of money in order to support rational investment decisions, which is known as the discounted cash flow method or net present value accounting. The net present value of an investment is equivalent to the projected net cash flows during a particular project, which are then capitalized to the actual moment of decision-making, while using the expected cost of capital or interest rate in the discounting procedure. To be specific, capitalization, as one of economic science's innovations, has been in use since the early fourteenth century by Dutch-Belgian and Italian merchant cities in the making of their calculations on trade and commercial credit. Nevertheless, the technique has effectively entered mainstream corporate finance theory since the 1950s (Faulhaber and Baumol, 1988).

Modern finance has taken the discounted cash flow model to a higher level of sophistication through the incorporation of uncertainty and risk and by fleshing out its implications for the estimation of the correct interest rate to be used in the capitalization and discounting procedures (Nitzan and Bichler, 2009). The mainstream ideas on risk have been disseminated by the work of Black and Scholes (1973) and Markowitz (1959) on the capital asset pricing model and portfolio theory, respectively. The basic message is that the effective interest rate to be used in the capitalization procedure is composed of a risk-free rate and a risk-related "mark-up", or premium, for specific investments. The latter is correlated with the fluctuation in returns of specific investments as compared with the average remuneration on a hypothetical market portfolio (as represented by the market security line). In a way, modern finance theory generates a moral correlation between the risk profile of an investment portfolio (as measured by the volatility of return on investment) and its return (Nitzan and Bichler, 2009).

There are a number of serious problems with the approach of mainstream finance in calculating risk and finding the right interest rate in the discounting and capitalization of a projected stream of revenues. The first is a logical one, reinforced by empirical evidence. As argued by Nitzan and Bichler (2009) in their analysis of the political economy of capitalization, "really existing" capitalists are actually keen on breaking this moral link between volatility and earnings growth, considering that reducing earning volatility is crucial for maintaining profitability. One of their examples relates to the trajectory of large US corporations over the last half century, according to which the earnings of a company like General Electric (GE) rose ten times faster than those of General Motors (GM), though the volatility of GE's earnings growth was far smaller than GM's. A second problem is that in a scenario of relatively thin and undeveloped capital markets there is no easy way to make the calculations that underpin the capital asset pricing model of modern finance theory. To be specific, in the absence of deep and consolidated capital markets, there is no "average portfolio" that performs the role of a benchmark in order to guide the calculations regarding the volatility of individual asset returns. This is particularly problematic in the context of emerging capital markets of the Global South in general, and in relation to cities as potentially cumbersome and uncooperative asset classes in particular, considering their lack of liquidity, standardization and commensurability. In short, considering these weaknesses of modern finance theory, the capitalization of cities as a potential asset class, particularly in the Global South, is in need of a stronger conceptual basis to explain risk dimensions in general and the choice of the interest rates in the discounting procedure in particular.

In that sense, it is worthwhile to recall our previous discussion on the state's absorption and sharing of risk in the making of markets for credit and finance. On the one hand, from the perspective of potential investors, cities in the Global South concentrate enormous deficits in terms of the supply and maintenance of infrastructure, housing and environmental services, which, at least theoretically, imply opportunities. On the other, the significant legal-institutional uncertainties that surround the demarcation of assets (assetization) and the design of associated revenue streams, as well as the relative lack of comparable program experiences, all represent challenges for the elaboration and circulation of standardized metrics and calculation practices. Thus, while cities are cash-strapped and badly in need of investments to live up to their growth potential, the initial systemic risks of building up and consolidating emerging capital markets are unlikely to be taken up by financial actors and investment bankers alone.

As we have discussed earlier on, this impasse is resolved through the transformation of the public fund, as anti-value, in state-mediated "fictitious value", which is operationalized through the rolling out of favourable regulatory frameworks at the national and local scales, the provision of grants and subsidies that allocate liquidity into cities and metropolitan areas, as well as the design of public guarantees linked to assets such as land, infrastructure and tax receivables, among some of the more frequent examples. This market-enabling absorption of risk also affects the interest rates that are used in the discounting of projected periodic income streams, and raises the net present value of assets. In other words, turning Christopher's (2018) analysis on risking value theory upside down, the state is directly involved in the valuation and pricing of risk, thereby contributing to asset price escalation or, in Lapavitsas' (2013) terminology, in the rise of fictitious capital.[13]

In order to better explain the mechanisms at work, in Table 2.1 we provide a simple numerical illustration of differential capitalization, that is, the effect of

Table 2.1 Risk, differential capitalization and fictitious capital

A - Assumptions				
(1) Risk-free rate	3%			
(2) Initial Risk premium	3%			
(3) Initial Cost of capital = (1)+(2)	6%			
(4) State driven adjusted risk premium (in year 2)	1%			
(5) Adjusted cost of capital = (1)+(4)	4%			
(6) Projected yearly revenue stream	R$500,000			
B - Asset Prices	*Net Present Value*	*Projected yearly income stream*		
		Year		
	(Year 1)	*1*	*2*	*3*
(7) Initial Capitalization (*)	**R$1,416,696.33**	R$500,000	R$500,000	R$500,000
(8) Differential capitalization (**)	**R$1,443,047.34**	R$500,000	R$500,000	R$500,000
(9) Fictitious Value of state-driven risk reduction = (8)–(7)	**R$26,351.00**			

Source: Author's elaboration.

(*) R$ 1,416,696.33 = R$ 500,000 + R$ 500,000/(1,06) + R$ 500,000/(1,06)^2
(**) R$ 1,443,047.34 = R$ 500,000 + R$ 500,000/(1,04) + R$ 500,000/(1,04)^2

the use of the public fund and state-driven gradual reduction of risk and interest rates on increasing net present values of assets.[14]

The example considers an asset that is supposed to generate a contracted yearly income stream of R$ 500,000 during three years. Moreover, given an initial scenario marked by a relatively high-risk premium of 3% (on top of the risk-free rate of 3%, which generates a total cost of capital of 6%), the asset is priced at R$ 1,416,963. A state-driven strengthening of the legal-institutional framework that guides private investment decisions is assumed to reduce the risk premium to 1% from the second year onward (generating a lower risk-adjusted cost of capital of 4%). As can be seen, this triggers an increase in the price of the same asset to R$ 1,443,047.34.

It is worthwhile to reflect on the significance of this asset price escalation, or fictitious value creation, of R$ 26,351. In a way, the quantitative result reflects the use of the public fund in the absorption and management of systemic risks, which are inherent to a scenario marked by the exploration and constitution of new markets. Considering that these risks are unlikely to be assumed by private actors, the state is effectively involved in the restructuring of the public fund itself – ex-ante anti-value under spatial Keynesianism – into a "collective agencement", through the circulation of norms and expectations aimed at the creation of "value for money" through the pricing/valuation of risk in the absence of consolidated credit markets. In the empirical part of the book we will provide illustrations of how the rollout of credit and capital markets is entangled with the mobilization of the state at multiple scales, its urban planning framework and the associated financial and legal devices aimed at the reduction of risk through valuation and pricing practices, outright subsidization and the provision of public guarantees.

The state-driven initial escalation of asset prices and fictitious value, as illustrated in a hypothetical manner in Table 2.1, also provides a favourable platform for the generation of successive rounds of financial profits through gearing or financial leverage. Hilferding (1910) had already observed this process in the beginning of the twentieth century, when relativizing the conflicts and pointing out the complementarities between finance and industrial capital during the emergence and growth of the listed corporation. More specifically, whenever the profit rate exceeds the interest on borrowing, industrialists could increase profitability by augmenting the ratio of debt to equity.[15]

Hilferding's (1910) conceptualization of "founder's" or "promoter's" profit as a driver behind financial profit in case of floating and commercialization of stock in capital markets is also related to the differential rates used in discounting procedures. To illustrate, when an enterprise floats its shares in order to raise capital, the emission value of its stock is the capitalization of the expected future dividends (for simplicity, assuming no debt finance and all profits paid out in dividends). A key issue here is the discount factor to be used in the capitalization procedure. If we assume that interest rates (for example 4%) are lower than profits (for example 6%),[16] and that the hypothetical enterprise has an infinite life generating a yearly profit (and dividend) of R$ 500,000, we end up with a similar numerical scenario that was illustrated in Table 2.1 (the only difference being the annuity of R$ 500,000 now being ad-infinitum). Under these conditions, the floating of the company's shares would generate a founder's profit to the initial owner and the intermediaries that facilitated the share issue of R\$ 4.2 million.[17] In other words, the liquidity (or loanable funds) of the buyers of the shares generate(s) an increase of fictitious capital, considering the cash of R\$ 12.5 million represents an advance payment for an ad-infinitum contracted income stream of dividends of R\$ 500,000, which is discounted with a lower rate of 4%. In favourable institutional and financial environments marked by gradually falling interest rates and higher profit expectations, the projection of founder's profit is likely to trigger successive rounds of trading and speculation of "coupons" in secondary markets and further increases in asset prices and fictitious value. Moreover, in this Minsky-inspired speculative environment of rising optimism and exuberance in relation to the expected rate of profitability, "good" investment projects, (i.e. those "considered" capable of earning an internal rate of return that exceeds interest rates), tend to push the economy rather than the availability of savings as such.[18]

A third and last dimension in our political economy of valuation is that the generation of founder's or promoter's profit in the urban setting requires an "upgrade" and transformation of the discounted cash flow and capitalization models from corporate finance, from a rather fragile "description" of asset pricing in efficient markets, into a device that can be put to work to transform city space into a bundle of contracted income streams. A political economy of valuation (Greco, 2015) is able to provide the earlier discussed literature on performativity with a structural grounding, which enables it to absorb conflict, ideology and power. In order to do so, however, we need what Jamie Peck (2017a, 2017b) has called mid-level concepts that allow an articulation between specific experiences and broader theory.

"Performative urbanism" represents such a mid-level concept. It provides a methodological bridge between the broader ambitions of urban political economy and the context-specific cases through which the state and urban planning are mobilized in the making of founder's profit in the city (Sanfelici and Halbert, 2018).

In relation to the specific Brazilian scenario, it should be observed that planning has traditionally performed an ideological role. For the Brazilian readership, this is nothing new considering the path-breaking work of authors such as Maricato (1996) and Villaça (1999) on the ideological character of master planning during the populist and authoritarian phases of the Brazilian development trajectory throughout much of the twentieth century (see also Chapter 3). These authors were pioneers in pointing out, in an interpretation that was radically different from the traditional normative analysis regarding the lack of effectiveness of planning in third-world cities, that Brazilian master plans were "performing" with an extraordinary degree of effectiveness their role as smokescreens. Plans performed the role of not revealing to society the contradictory and selective social (re)production of space in cities and metropolitan areas. More recent planning achievements have also been slow to change this structural feature of Brazilian cities. The widely acclaimed national framework legislation of the City Statute on urban reform and participatory-collaborative planning was approved in 2001. It enabled local governments to design and implement a series of redistributive land use instruments aimed at combatting speculation and the provision of well-located and accessible low-income housing. Nevertheless, as will also be discussed in the subsequent chapters, it has faced considerable challenges to live up to the initially high expectations to change the praxis in Brazilian cities (Klink and Denaldi, 2014).

The previously mentioned critical literature was path-breaking in its own right for spelling out the peculiar grammar of the informality, socio-spatial and racial exclusion and environmental degradation that underpinned the functioning of Brazilian cities. However, it has not dealt with the increasing articulations between the urban planning framework and finance itself. We will argue that the structuring dimension of the plan as ideology is now gradually being reinforced by financial modelling as a complementary representation of space in times of financialization. This is the essence of what we will call "performative urbanism", whereby financial models not only describe the projected returns of transforming city space into assets but actually contribute to the state-mediated making of founder's profit. In a way, "performative urbanism" is instrumental in the "urbanization of finance" itself in the sense of articulating the linkages between apparently contradictory representations of space, (i.e. as tradable income yielding assets versus the privileged domain of use value and the reproduction of life).

An investigation of "performative urbanism" and how (and why) planners, financial intermediaries and other actors are involved in putting the discounted cash flow modelling to work in the transformation of cities also requires fleshing out the specific calculative practices and institutional-legal devices that are mobilized in these processes. This means tracing the politically contested, social delimitation of specific assets and associated income streams, investigating the distribution of risks between the state and private partners and creating a deeper understanding of the narratives that are used to fill in the specific parameters of the models that are being disseminated in order to guide the planning, management and finance of Brazilian cities. In relation to the latter, the approximation between modelling and planning has, not coincidently, been accompanied with a number of narratives on the potential transformative praxis of finance in cities. In a way, these narratives serve to reinforce the relational roles of finance capital and planners in transforming the city's "negative externalities" into development potential, evidently under the condition that the right parameters are filled into the financial equations. In this book we will discuss several examples of these "representational spaces of finance", such as the hypothetical alleviation of public budgets by unlocking the potential of capital markets, the mobilization of professional management and know-how that is disciplined by the driving force of shareholder value premises, or the creation of pseudo-markets that enable to get urban prices right, target more efficiently the limited available subsidies as well as eliminate the inefficient use of scarce environmental resources.

The empirical chapters in this book will provide more contemporary evidence of these interactions between planners, financial actors and other intermediaries in the pricing and valuation of debt, public assets and infrastructure networks in putting the discounted cash flow model to work and "urbanize finance" in order to transform the Brazilian metropolis into a contracted stream of revenue and rents. In each of these examples, we will show the role of the language of financial modelling in structuring the conversations between urban planners and other actors. The experiences show that, in a way, the variations of discounted cash flow models and related financial metrics have provided new entrance points for financial and non-financial actors alike, for the gradual filling in of urban policies aligned with their specific projects and strategies. While this tends to "de-contextualize" urban planning from its official, more traditional land-use and zoning narratives (Savini and Aalbers, 2016), the analysis of the experiences at the same time illustrates that the renewed penetration of finance capital in Brazilian cities is at best still open-ended and contested.

Finally, it should be stressed that, considering its history as a rural, colonial and export-led economy, the multiple entanglements between Brazil, its city-network and international finance is not a particularly new phenomenon. As will be discussed in Chapter 3, before the definite take off of its industrialization and urbanization, both domestic loanable funds and global finance prioritized Brazilian cities as emerging spaces of urban financial profits and rents.

2.3   From nationally scaled dependent-subordinated to multi-scalar interdependent financialization?

As mentioned, the bulk of the literature that makes reference to the Global South has been structured around the idea of dependent-subordinated financialization (Lapavitsas, 2013; Paulani, 2011). This is understandable, particularly in light of the fact that the concept echoes earlier debates and circulation of ideas on centre-periphery and dependency theory. This occurred from the early-mid twentieth century onwards and was triggered by intellectuals such as Celso Furtado, Myrdal and Prebish and key institutions like the United Nations Economic Commission for Latin America and the Caribbean (CEPAL), among others. The centre-periphery model projected deteriorating terms of trade (i.e. the relation between export and import prices) for the Global South, which was related to their export specialization in unprocessed agricultural and resource intensive goods, while importing manufactured final products from the developed world. Without effective state intervention in this laissez-faire international order of trade and commerce, the periphery was bound to fall behind the pattern of growth and development that was occurring in the centre. The centre-periphery metaphor proved to be a powerful argument in the design and implementation of national strategies and projects structured around import substitution and industrialization (Bielschowsky, 1988).

The recent work on dependent-subordinated financialization revisits this opposition between developed market economies and the global periphery but provides it with a theoretical framework that is structured around a hierarchical monetary and financial order that is dominated by the centre. In most of this literature, the role of the US dollar as the main international reserve currency represents a key message. In the prevailing international order, which is dominated by the US as the emissary of the informal world money, countries of the Global South are once again situated in the periphery, considering their lack of a convertible national currency. Consequently, the South faces a large volume of globally circulating short-term financial capital, which forces them to hold a significant amount of dollar reserves as a mechanism of "self-insurance" aimed at reducing country risks. This massive holding of international dollar reserves represents "money flowing uphill" from the periphery to the centre, considering that savings from the Global South absorb the current account deficits from the US. Moreover, domestic holdings of dollar reserves are costly, considering these pay low interest when compared with the domestic rates that could have been obtained.[19] Table 2.2 and 2.3 provide an overview of the significant and growing amounts of reserves held by the Global South as a whole and by selected countries in particular.[20] As

Table 2.2 Emerging and developing economies currency composition of official foreign exchange reserves (US Dollars, millions)

	2001Q1	2011Q1	2015Q1
Total Foreign Exchange Reserves	726,532.46	6,505,977.79	7,486,757.89
Allocated Reserves	416,710.71	2,544,696.69	3,219,404.69
Unallocated Reserves	309,821.75	3,961,281.10	4,267,353.19
Shares of Allocated Reserves (%)	57.36	39.11	43.00
Shares of US dollars	75.20	59.03	67.75
Shares of euro	18.30	27.35	18.29
Shares of Japanese yen	2.31	2.86	2.84
Shares of pounds sterling	2.63	5.79	4.63
Shares of Australian dollars			1.59
Shares of Canadian dollars			1.89
Shares of Swiss francs	0.19	0.06	0.06
Shares of other currencies	1.37	4.90	2.95
Shares of Unallocated Reserves (%)	42.64	60.89	57.00

Source: Author's elaboration. Based on: IMF: Currency Composition of Official Foreign Exchange Reserves (COFER), International Financial Statistics.

Starting 2015Q2, the breakdown for emerging and developing economies is no longer available.

Data available at: http://data.imf.org/ on: 5/4/2019 10:03:12 AM. (Accessed August 14, 2019)

Table 2.3 International reserves and foreign currency liquidity: Selected countries 2001, 2011 and 2019. Stock (Billion USD) and yearly annual growth

	2001	2011	Average annual growth 2001/11 (%)	2019	Average annual growth 2011/19 (%)
Brazil	35.60	352.00	25.75%	384.17	1.10%
India	46.40	262.90	18.94%	412.87	5.80%
Russia	33.10	454.00	29.93%	487.80	0.90%
China	216.30	3,181.10	30.84%	3,090.18	−0.36%
Mexico	44.80	142.50	12.27%	168.87	2.15%

Source: Author's elaboration. Based on the International Monetary Fund: "Currency composition of Official Foreign Exchange Reserves (COFER)" Available at: http://data.imf.org/?sk=E6A5F467-C14B-4AA8-9F6D-5A09EC4E62A4&sId=1408243036575 (Accessed: April 4th, 2019).

illustrated in Table 2.2, the holding of foreign exchange reserves has multiplied by more than 10 times during 2001–2015, that is, increasing from 726 billion USD in 2001 to almost 7.5 trillion USD in 2015. At the same time, this growth has flattened significantly after 2011, while China has started to reduce its portfolio of currency reserves. While the share of the USD in the overall portfolio of currency holdings of emerging and developing economies is around two-thirds, it tends to fluctuate in times of crisis. Table 2.2 also indicates that the US subprime crisis has triggered a move away from the dollar, resulting in a reduction of more than 15 percentage points of the US dollar's share in overall holdings of currency reserves of emerging and developing economies between 2001 and 2011.

The notion of dependent-subordinate financialization structured around the idea of money flowing "upward" from peripheral countries to the centre in light of the dollar's role as informal world money is problematic for several reasons.

First, it is challenging to analyze the prevailing, undeniable hierarchical financial and monetary world order in terms of flows between homogenous groupings of peripheral and central economies.[21] Table 2.4 provides a snapshot of the main net lending and borrowing positions of the advanced economies as well as the emerging and developing economies during the period 2001–2018, organized according to the country classification of the International Monetary Fund's World Economic Outlook.[22] To be clear, positive figures on net lending reflect the sum of the current account (the difference of national domestic savings minus investments) and the capital account, indicating an outflow of domestic savings from specific countries in order to finance net borrowing of the rest of the world.

Although for the emerging market and developing economies group as a whole, money is flowing uphill during 2001–2012, since 2015 net lending and borrowing more or less cancel out. Moreover, when we open up for specific sub-regions, the picture becomes more complex. For example, emerging and developing Europe and Latin America and the Caribbean have been consistent net borrowers from the rest of the world during the period considered. Likewise, the net lending by poor Sub-Saharan Africa to the rest of the world, which was detected by Lapavitsas (2013) in his analysis on dependent financialization, has since 2013 changed into a flow of savings into the region. For example, net lending from the rest of the world in 2015 represented 5.5% of the region's Gross domestic Product. The same is true in relation to the figures on net lending and borrowing of advanced countries. While during the period 2001–2012 this group has performed the role of net borrower, since 2013 these economies transformed into net lenders. To be specific, Table 2.4 shows that during the period 2013–2018, advanced countries were lending around 0.5% to 0.7% of their GDP to the rest of the world. Likewise, opening up this very diverse set of economies shows significant discrepancies. While the US has indeed depended on foreign borrowing in light of consistent current account deficits, particularly during the period 2013–2018 the Euro Area as a whole has lent between 2.4% and 3.1% of its GDP. Japan shows similar figures, with a savings surplus that results in a net lending to the rest of the world of around 3.5% of its GDP in 2018. In short, these numbers suggest that money is both flowing uphill from and downhill to the global periphery.

Table 2.4 Summary of net lending and borrowing (Percentage of GDP)

	Averages							
	2001–10	2005–12	2013	2014	2015	2016	2017	2018
Advances Economies								
Net Lending and Borrowing	-0.7	-0.6	0.5	0.5	0.6	0.7	0.8	0.7
Current Account Balance	-0.8	-0.6	0.5	0.5	0.6	0.7	0.9	0.7
Savings	21.6	21.4	21.9	22.5	22.7	22.2	22.7	22.7
Investments	22.3	22.0	21.1	21.4	21.5	21.3	21.6	21.9
Capital Account Balance	0.0	0.0	0.0	0.0	0.0	0.0	0.0	0.0
USA								
Net Lending and Borrowing	-4.4	-4.0	-2.1	-2.1	-2.2	-2.3	-2.2	-2.3
Current Account Balance	-4.4	-4.0	-2.1	-2.1	-2.2	-2.3	-2.3	-2.3
Savings	17.3	16.8	19.2	20.3	20.1	18.6	18.9	19.0
Investments	21.5	20.8	20.4	20.8	21.0	20.3	20.6	21.1
Capital Account Balance	0.0	0.0	0.0	0.0	0.0	0.0	0.0	0.0
Euro Asia								
Net Lending and Borrowing	0.1	0.1	2.4	2.6	3.1	3.2	3.1	3.1
Current Account Balance	0.0	0.0	2.3	2.5	2.9	3.2	3.2	3.0
Savings	22.7	22.6	22.4	22.9	23.7	24.2	24.7	24.8
Investments	22.3	22.0	19.6	19.9	20.4	20.8	20.9	21.5
Capital Account Balance	0.1	0.1	0.2	0.1	0.2	0.0	-0.2	0.1
Germany								
Net Lending and Borrowing	4.1	5.9	6.7	7.6	8.9	8.6	8.0	7.4
Current Account Balance	4.1	5.9	6.7	7.5	8.9	8.5	8.0	7.4
Savings	24.0	25.7	26.2	27.1	28.1	28.2	28.1	28.6
Investments	19.9	19.8	19.5	19.6	19.2	19.7	20.1	21.2
Capital Account Balance	0.0	0.0	0.0	0.1	0.0	0.1	0.0	0.0
France								
Net Lending and Borrowing	0.7	-0.3	-0.5	-1.0	-0.4	-0.7	-0.5	-0.6
Current Account Balance	0.7	-0.3	-0.5	-1.0	-0.4	-0.8	-0.6	-0.7

(Continued)

Table 2.4 (Continued)

	Averages							
	2001–10	2005–12	2013	2014	2015	2016	2017	2018
Savings	23.0	22.5	21.8	21.8	22.3	21.9	22.9	22.1
Investments	22.4	22.9	22.3	22.7	22.7	22.7	23.5	22.8
Capital Account Balance	0.0	0.0	0.0	-0.1	0.0	0.1	0.0	0.1
Italy								
Net Lending and Borrowing	-1.2	-1.8	0.9	2.1	1.7	2.4	2.7	2.5
Current Account Balance	-1.3	-1.9	1.0	1.9	1.5	2.5	2.8	2.6
Savings	19.9	18.7	17.9	18.9	18.8	20.1	20.4	20.6
Investments	21.1	20.7	17.0	17.0	17.3	17.6	17.6	18.0
Capital Account Balance	0.1	0.1	0.0	0.2	0.2	-0.2	-0.1	-0.1
Spain								
Net Lending and Borrowing	-5.5	-5.4	2.2	1.6	1.8	2.5	2.1	1.0
Current Account Balance	-6.1	-5.9	1.5	1.1	1.2	2.3	1.8	0.8
Savings	21.9	20.7	20.2	20.5	21.6	22.7	22.9	22.7
Investments	28.0	26.5	18.7	19.5	20.4	20.4	21.1	21.9
Capital Account Balance	0.6	0.5	0.6	0.5	0.7	0.2	0.2	0.3
Japan								
Net Lending and Borrowing	3.2	3.0	0.7	0.7	3.1	3.8	4.0	3.5
Current Account Balance	3.3	3.1	0.9	0.8	3.1	4.0	4.0	3.5
Savings	27.4	26.3	24.1	24.7	27.1	27.4	27.9	27.9
Investments	24.1	23.2	23.2	23.9	24.0	23.4	23.9	24.4
Capital Account Balance	-0.1	-0.1	-0.1	0.0	-0.1	-0.1	-0.1	0.0
United Kingdom								
Net Lending and Borrowing	-2.9	-3.2	-5.2	-5.0	-5.0	-5.3	-3.4	-4.0
Current Account Balance	-2.9	-3.2	-5.1	-4.9	-4.9	-5.2	-3.3	-3.9
Savings	14.3	13.4	11.1	12.3	12.3	12.0	13.9	13.3
Investments	17.2	16.6	16.2	17.3	17.2	17.3	17.2	17.2
Capital Account Balance	0.0	0.0	-0.1	-0.1	-0.1	-0.1	-0.1	-0.1

Canada								
Net Lending and Borrowing	0.4	-1.1	-3.2	-2.4	-3.5	-3.2	-2.8	-2.6
Current Account Balance	0.5	-1.1	-3.2	-2.4	-3.5	-3.2	-2.8	-2.6
Savings	22.6	22.5	21.7	22.5	20.3	19.7	20.7	20.4
Investments	22.1	23.6	24.9	24.9	23.8	22.9	23.5	23.0
Capital Account Balance	0.0	0.0	0.0	0.0	0.0	0.0	0.0	0.0
Other Advanced Economies								
Net Lending and Borrowing	4.0	4.0	5.1	5.0	5.3	5.4	4.5	4.7
Current Account Balance	4.0	4.0	5.0	5.2	5.7	5.3	4.7	4.7
Savings	29.8	30.3	30.3	30.5	30.8	30.1	30.1	29.9
Investments	25.5	26.0	25.1	25.2	24.8	24.7	25.2	25.2
Capital Account Balance	0.0	0.0	0.1	-0.1	-0.4	0.1	-0.2	0.0
Emerging Market and Developing Economies								
Net Lending and Borrowing	2.5	2.7	0.7	0.6	0.0	-0.2	0.0	0.0
Current Account Balance	2.5	2.6	0.6	0.6	-0.2	-0.3	0.0	-0.1
Savings	30.3	32.6	32.8	33.0	32.4	31.9	32.2	32.7
Investments	28.1	30.3	32.4	32.6	32.8	32.1	32.4	32.8
Capital Account Balance	0.1	0.2	0.1	0.0	0.1	0.1	0.1	0.1
Regional Groupings:								
Commonwealth of Ind. States								
Net Lending and Borrowing	4.9	4.2	0.6	0.6	2.8	0.0	1.0	5.0
Current Account Balance	5.6	4.5	0.6	2.1	2.8	0.0	1.0	5.0
Savings	26.7	26.8	24.2	25.0	25.9	24.4	25.4	29.0
Investments	21.2	22.4	23.5	22.8	22.7	23.9	24.3	24.0
Capital Account Balance	-0.7	-0.3	0.0	-1.5	0.0	0.0	0.0	0.0
Emerging and Developing Asia								
Net Lending and Borrowing	3.7	3.7	0.8	1.5	2.0	1.4	0.9	-0.1
Current Account Balance	3.6	3.6	0.7	1.5	2.0	1.4	0.9	-0.1
Savings	39.7	43.0	43.1	43.6	42.5	41.1	41.0	40.1
Investments	36.4	39.6	42.3	42.1	40.0	39.7	40.1	40.2
Capital Account Balance	0.1	0.1	0.1	0.0	0.0	0.0	0.0	0.0

(Continued)

Table 2.4 (Continued)

	Averages		2013	2014	2015	2016	2017	2018
	2001–10	2005–12						
Emerging and Developing Europe								
Net Lending and Borrowing	-4.5	-5.1	-2.5	-1.6	-0.7	-1.2	-1.9	-1.3
Current Account Balance	-4.8	-5.7	-3.6	-2.9	-2.0	-1.8	-2.5	-2.2
Savings	19.6	20.0	21.4	22.0	22.9	22.3	22.9	23.1
Investments	24.2	25.7	24.9	24.9	24.8	24.1	25.4	24.8
Capital Account Balance	0.4	0.6	1.1	1.3	1.3	0.6	0.6	1.0
Latin America and the Caribbean								
Net Lending and Borrowing	-0.1	-0.5	-2.8	-3.0	-3.2	-1.9	-1.4	-1.9
Current Account Balance	-0.2	-0.6	-2.8	-3.1	-3.2	-1.9	-1.4	-1.9
Savings	20.4	21.0	19.1	17.9	16.5	16.9	16.8	17.6
Investments	20.6	21.6	22.3	21.5	21.1	18.5	18.5	19.6
Capital Account Balance	0.1	0.1	0.1	0.0	0.0	0.0	0.0	0.0
Middle East, North Africa, Afghanistan, Pakistan								
Net Lending and Borrowing	7.8	10.0	10.0	6.3	-3.7	-3.7	-0.7	2.4
Current Account Balance	8.2	10.6	9.8	5.5	-4.0	-3.9	-0.6	2.3
Savings	35.0	37.9	36.1	32.9	24.8	24.6	26.7	29.3
Investments	27.7	28.2	26.1	27.0	28.7	28.1	27.5	27.1
Capital Account Balance	0.1	0.1	0.0	0.2	0.0	0.0	0.0	0.1
Sub-Saharan Africa								
Net Lending and Borrowing	1.8	2.0	-1.7	-3.2	-5.5	-3.0	-1.6	-2.2
Current Account Balance	0.5	0.5	-2.2	-3.6	-5.9	-3.7	-2.1	-2.6
Savings	20.7	21.6	19.4	19.2	17.2	18.0	18.9	17.8
Investments	20.7	21.5	21.6	22.5	22.7	21.2	21.0	20.2
Capital Account Balance	1.3	1.5	0.4	0.4	0.4	0.7	0.6	0.4

Author's elaboration. Based on IMF's World Economic Outlook. Statistical Appendix. Table A14

Second, the centre-periphery and dependency perspective not only fail to rec-ognize the significant differentiation within the Global South but also does not consider other spatial scales that are relevant to understanding contemporary financialization. To be specific, the implicit focus of the dependency and centre-periphery conceptualization is the container of the national space economy. As such, it by and large ignores the entanglements between city spaces and the creation and circulation of assets in international capital markets. However, as also remembered by authors such as Dymski (1998), the macroeconomic figures on current account balances and net lending-borrowing positions of individual countries only represent one structural, non-spatial dimension that mark the com-plex international flows of savings, liquidity and assets. An explicitly spatial and multi-scalar perspective would allow to further flesh out the role of big cities and metropolitan areas within national space economies, both in terms of extracting domestic wealth from smaller and less dynamic cities (which accumulate "trade deficits" with domestic metropolitan areas), and as privileged points of refer-ence for anchoring international savings and liquidity (in case of national current account deficits).[23] More particularly, a context marked by large inflows of inter-national and national liquidity (that is, through the accumulation of successive current account deficits with the rest of the world and the inflow of domestic wealth in light of trade surpluses with smaller cities) and a relative lack of alterna-tive investment opportunities in productive-industrial accumulation could set the stage for urban-driven speculative founder's profit and escalation of asset prices along the lines we discussed in the previous section 2.2. As we will argue in the next chapter, within the Brazilian development trajectory, cities and metropolitan spaces have been privileged platforms for the accumulation of urban rents and financial profits.

A third conceptual problem is related to the association between the role of the dollar as implicit world money and the notion of "financial dependency" of the Global South as such (on developed market economies in general and the US in particular). As mentioned, there are undeniable opportunity costs and burdens for the Global South associated with the holding of large international currency reserves. Nevertheless, the prevailing system of world monetary governance rep-resents a complex "interdependent" (albeit asymmetric and structurally unstable) system in light of the mutual entanglements between the US, other key currency countries and the rest of the world. At a global level, the massive holding of dollar reserves has had a deflationary effect on the world economy, considering these resources could have been used for productive investments. At the same time, however, the role of the dollar as world money has allowed the US to perform its role as a "buyer of last resort" in order to maintain global effective demand (defined as the sum of household consumption, investments from firms and gov-ernment expenditures). It has done so by running consistent fiscal and trade defi-cits with the rest of the world. While the US benefit from receiving cheap credit from lending countries (by exporting US treasury bills and similar low-yielding titles to the rest of the world), key-currency countries gradually transform into net importers of goods and services, net exporters of jobs and also face continuous threats of underemployment and lack of domestic aggregate demand. There are inherent domestic political limits to such a trajectory. To aggravate this scenario, particularly in the twenty-first century, the role of the US dollar as "shadow" world money has been increasingly questioned in light of its fluctuations in relation to other currencies such as the euro, the yen and others. This has undermined the financial rationale of sticking to the dollar as the only international reserve currency.

The potentiality of concepts such as "mutual hostage" and asymmetric interdependency in trade, finance and exchange rate policy (Stiglitz, 2007: 258), as opposed to a twentieth century neo-industrial conceptualization in terms of structural dependency, is illustrated paradigmatically when looking at the recent bilateral relations between the USA and China. In 2005, the latter officially declared it would not accumulate dollar reserves anymore. At the same time, together with countries such as Japan – which had built up similar stocks of dollars – China was constrained in the massive selling of these assets in order to avoid financial losses on its remaining dollar reserves.[24] In the meantime, the US argued that its trade and current account deficits with China, resulting from a flood of cheap imports and leading to underutilized domestic industrial capacity and unemployment, were largely due to the latter's depreciated exchange rate. As a result, the US has been consistently pressuring the Chinese to reevaluate its currency. Nevertheless, this position clearly underestimated the structural feature of the US fiscal and trade deficits; a Chinese currency revaluation leads to a trade-shift in US imports from China to other competitive exporters, as reflected in the US recent trade patterns with Southeast Asia. Moreover, an increasing number of central bankers started to realize that a concentration on the dollar as the prime international reserve asset represented an unnecessary risk, particularly in light of its fluctuations. China, too, was aware that diversification of its portfolio of international assets not only made good sense but also signalled it was not "dependent" on the US in the sense of sending its exports goods and lending cheap money in exchange. As matter of fact, it could be argued that it is easier for China to channel its savings into domestic investments than it is for the US to find similar large-volume buyers of its US treasury bonds (Stiglitz, 2007: 259–260).

There is a fourth problem with prevailing formulations of dependent-subordinate financialization as money flowing uphill from the periphery to the centre where key reserve currencies and world money are emitted. Even recognizing a hypothetical correlation between a country's peripheral position in the world financial order and its net outflow of funds (i.e. representing its lending to the rest of the world),[25] there is no inherent link between the direction of the flow of savings and the (in)subordinate character of financialization in the Global South in the first place. For instance, while goods and savings have been flowing uphill from China to the US, the relation between these countries can be labelled as multiple dependence and lock-in. At the same time, a country such as Brazil has been a consistent net borrower. Table 2.5 shows figures on the current account, which are representative of the country's net savings position in relation to the rest of the world. With the notable exception of the years between 2003–2007, Brazil's

Table 2.5 Brazilian current account balance: US dollars (billions) and as percentage of GDP

	US Dollars, Billions	*Percentage of GDP*
1993	−0.59	−0.14
1994	−1.68	−0.32
1995	−18.71	−2.38
1996	−23.84	−2.80
1997	−30.85	−3.49
1998	−33.89	−3.92
1999	−25.87	−4.31
2000	−24.79	−3.78
2001	−23.72	−4.24
2002	−8.10	−1.59
2003	3.76	0.67
2004	11.35	1.70
2005	13.55	1.52
2006	13.03	1.18
2007	0.41	0.03
2008	−30.64	−1.81
2009	−26.26	−1.58
2010	−79.01	−3.58
2011	−76.29	−2.92
2012	−83.80	−3.40
2013	−79.79	−3.23
2014	−101.43	−4.13
2015	−54.47	−3.03
2016	−24.01	−1.34
2017	−7.24	−0.35

Source: Author's elaboration. Based on International Monetary Fund, World Economic Outlook Database, April 2019.

structural domestic savings deficit (or negative current account) indicates that money has been consistently flowing "downhill" from the rest of the world to the country from 1993 to 2017.

Nevertheless, during the same period, it can be argued that Brazil has walked a line of subordinated financialization, particularly in light of its failure to design and operate a domestic market for shares, bonds and long-term credit aimed at the finance of its urbanization and industrialization (Tavares, 1999), in spite of the availability of loanable funds.

In the next chapter we will flesh out this argument by looking at the Brazilian development trajectory since the early twentieth century and its entanglements with city space. The country's industrialization strategy was, by and large, financed by a combination of active developmental state intervention, international capital and, particularly in the case of sheltered sectors with a significant presence of family groups such as building and construction, reinvested profits. Thus, the country's massive transformations were implemented in spite of the absence of consolidated and liquid domestic capital markets for long-term credit and shares. Instead, a significant part of the available domestic loanable funds that were built up during the Brazilian development trajectory were instrumental in providing the speculative urban founder's profits associated with the highly selective production and finance of the country's emerging national network of big cities and metropolitan areas, which were accumulating huge socio-spatial and environmental deficits. As a matter of fact, when analyzing the monetary stabilization of the *Plano Real* and the subsequent re-insertion of the Brazilian economy into the global financial economy during the 1990s, Brazilian scholar Fiori (1997) sarcastically argued that the Plano Real had once again exposed the Brazilian elite's historically more modest ambitions as a minor player within the re-emerging financialized international order. In his view, the flow of international savings was never really meant to finance a new cycle of investments, which would be grounded within a broader national development project. Instead, it merely reflected what he called an option of subordinated insertion into the international financial order orchestrated by a "Brazilian buying and *rentier* bourgeoisie". It would have been national investments rather than (national or international) savings that could have pulled the Brazilian economy out of its transitional impasse that marked much of the 1990s. Nevertheless, with yearly real interest rates around 40% in the initial years of the Plano Real, compounded by uncertainties that surrounded the newly constituted currency and a history marked by decades of very high inflation, this was unlikely to generate a scenario whereby potentially productive investment projects would pass the test of credit rationing of bankers and finance capital.[26]

In short, neither the massive and costly holdings of international dollar reserves, nor the direction of the international flow of savings as such, explain what is called subordinate financialization in Brazil. Rather, this accumulation of dollars and the continuous dependence on external borrowing are not the cause but a consequence of the lack of broader national development project that could have mobilized its domestic loanable funds around a sustainable trajectory of industrialization and urban justice. The consistent inflow of international savings and domestic wealth, without having triggered sustainable investments, growth and social inclusion in big cities and metropolitan areas, is a direct reflection of this vacuum, as well as the lack of a strategy in relation to the country's insertion into the post-Bretton Woods international monetary order. Instead, without the obvious benefit of holding a convertible currency, which would have enabled it to export part of the accumulated tensions to the rest of the world (although, as we have seen, even for the US this process has limits), the Brazilian trajectory has been marked by a continuous state rescaling and restructuring in order to alleviate and manage financial-monetary tensions and crises within its national space economy, with direct implications for the highly selective finance of urban and metropolitan areas, as well as the privileged role of cities in the generation of founder's profits and the (early) urbanization of finance. In the next chapter we provide a broader historical analysis on how this contradictory pattern has unfolded, including an evaluation of the role of monetary governance and the state in the finance of cities. It will also set the stage for the case studies in the rest of the book.

Notes

1 As also mentioned in the previous chapter, this critique had been raised in relation to French regulation school's reading of the transformation from Fordism to post-Fordism (Jessop, 2000).

2 For instance, Barnes' (1996) perspective on urban geography was aimed at the elaboration of what he called "post pre-fixed" theory.

3 Planning as a "praxis" is defined in terms of "the normal everyday attempts by social collectivities to develop and manage their life worlds, building, living, thinking" (Law-Yone, 2007: 318).

4 The research undertaken by authors such as Peck and Whiteside (2016) on austerity, restructuring and rescaling of the city-capital market nexus and its impacts on financialized urban governance represents an exception to this pattern in the literature.

5 In a way, our approach goes into the same direction as authors such as Dymski (2009: 439) when he recommends that: "those involved in understanding the urban problematic make deep contact with heterodox economists".

6 In the next chapter we will discuss specific examples of how commodity money was used in nineteenth-century Brazil under conditions of scarcity of gold and other forms of hard cash or liquidity. While coffee was instrumental as a means of payments, slaves represented a mechanism to store and preserve value. See also Costa (2012: 426).

7 In the next section we will work out in more detail the concept of financial profit for the urban context.

8 It is outside the scope of this book, but the creation of intermediary-continental scales for monetary policy design and implementation, as reflected in the experience of the European Central Bank and the creation of the euro, is another example of this continuous search for a more balanced articulation between states, scales and territory in a scenario of imperfect global monetary governance.

9 Neoclassical economics and finance theory only focuses on "standardized time", where well-functioning and competitive markets exist, with actors assuming risks that are strictly correlated to the fluctuations of individual investment portfolios in relation to a hypothetical (and known) market portfolio. This excludes systemic risks associated with market making. This assumption allows "normal economic science" to move away from the complexities that surround the social constitution of money, credit and finance, as well as their preservation in times of imminent collapse.

10 In relation to the constitution of financialization, we will analyze large urban redevelopment projects financed with securitized building rights (Chapter 4), the penetration of shareholder value premises in collaborative state-municipal water governance (Chapter 5), the mobilization of the state in the design and making of "market friendly" social housing finance systems (Chapter 6), and the reframing and restructuring of the city-capital market nexus in times of crisis and austerity (Chapter 7), among some of the examples discussed. With regard to financial crisis management, Chapter 3 discusses historical examples of how the state has recurrently absorbed the economic costs of financial bankruptcies as well as stabilization programs.

11 As remembered by Birch (2017), assets are not synonymous to commodities in the sense that the latter can also be traded in markets but do *not* generate a periodic income associated with their ownership. In a way, this is the difference between buying a CD in the record shop (commodity) or downloading the same music by paying a user charge to the company that owns the copyrights of the song or artist (asset).

12 In Chapter 5 we develop a somewhat similar argument for water pricing and the making of financialization in the context of Metropolitan São Paulo.

13 According to Lapavitsas (2013), net present value can be interpreted in terms of Marx's conceptualization of fictitious capital. This is particularly relevant when analyzing the role of the state in the making of financialization through its influences on Net Present Value calculations made by communities of financial and non-financial actors alike.

14 It should be observed that our notion of differential capitalization as one of the quantitative dimensions of the absorption of risk by the state in the making of urban financialization is related but more specific than Nitzan and Bichler's (2009) conception. The latter analyze differential capitalization in terms of the asymmetric power relations that are involved in the design and implementation of the capitalization formula, both in terms of the numerator (hype and exaggeration in the estimation of projected revenue streams) as well as in the interest rates that are used in the discounting procedure. For Nitzan and Bichler, the essence of differential capitalization is power.

15 The principle of financial leverage is a key element of modern finance theory and also contributes to problematize the supposed inherent contradiction between industry and finance, a misunderstanding that continues to hunt some of the literature on financialization. See also Chapter 1.

16 As mentioned by Lapavitsas (2013), Marx adopted the same assumption.

17 This is calculated as the difference in present value of two equivalent ad-infinitum income streams that are capitalized with different interest rates. In other words, this amounts to the difference between R$ 500,000/4% = R$ 12.5 million and R$ 500,000/6%= R$ 8.3 million.

18 This is different from projects *effectively* earning an internal rate of return that is higher than the cost of capital. The difference becomes essential at the turning point of economic cycles, when bubbles tend to burst.

19 There are additional costs. For instance, in order to avoid inflationary pressure from the inflow of dollars, national central banks sterilize/buy the additional dollars through the sale of domestic debt titles in the local market, which imply a higher interest burden. Moreover, domestic capitalists in the global south have actively explored the role of the dollar reserves as self-insurance. To be specific, they have effectively used the interest rate spreads to implement "carry trade", that is, borrow internationally at low rates and reinvest in domestic financial assets with higher yields. This has imposed additional tributes and opportunity costs and has increased the disparities within countries of the Global South.

20 Starting 2015, Quarter 2, the breakup of the times series for emerging and developing countries is no longer available.

21 As a matter of fact, Lapavitsas (2013) does distinguish between two broad segments within the Global South. The first comprises a set of countries that have gained a significant share in the export markets of manufactured goods to the US and Europe (China being the main example). The second group is composed of countries that export commodities, oil and unprocessed mineral products (including Russia and the Gulf countries). Nevertheless, this broad classification fails to recognize the significant differences within the Global South in respect to the direction of the flow of funds to the rest of the world.

22 The specific classification of the IMF for emerging markets and developing economies is the following. Commonwealth of independent states: Azerbaijan, Kazakhstan, Russia, Turkmenistan, Tajikistan and Uzbekistan. Emerging and developing Asia: Brunei Darussalam, Timor-Leste, Kiribati, Lao P.D.R., Marshall Islands, Mongolia, Papua New Guinea, Solomon Islands and Tuvalu. Latin America and the Caribbean: Bolivia, Brazil, Ecuador, Trinidad and Tobago, Venezuela, Argentina, Chile, Guyana, Paraguay, Peru, Suriname and Uruguay. Middle East, North Africa, Afghanistan and Pakistan: Algeria, Bahrain, Iran, Iraq, Kuwait, Libya, Oman, Qatar, Saudi Arabia, United Arab Emirates, Yemen, Afghanistan, Pakistan, Mauritania, Somalia and Sudan and Sub-Saharan Africa: Angola, Chad, Republic of Congo, Equatorial Guinea, Gabon, Nigeria, South Sudan, Burkina Faso, Burundi, Central African Republic, Democratic Republic of the Congo, Côte d'Ivoire, Eritrea, Guinea, Guinea-Bissau, Liberia, Malawi, Mali, Sierra Leone, South Africa, Zambia and Zimbabwe. The data have been collected from the IMF "World Economic Outlook 2019". Statistical Appendix. A14.

23 The trade deficit accumulated by less dynamic, smaller cities with big cities and metropolitan areas of the same national space economy leads to a net outflow of funds and wealth from the former to the later. In a way, this is similar to the flow of funds pattern that was prevailing among countries under the gold standard (considering that cities and metropolitan areas do not control exchange and monetary policies).

24 Table 2.3 shows that China's stock of international reserves has indeed been reduced very gradually during the period between 2011 and 2019 (a yearly reduction of 0.36%).

25 Our argument was that there isn't.

26 Evidently, as will be seen in the case studies, profit-generating projects are not necessarily equivalent to equitable and/or sustainable projects.

References

Aglietta M (2018) *Money: 5,000 years of debt and power*. London and New York: Verso.

Barnes TJ (1996) *Logics of dislocation: Models, metaphors, and meanings of economic space*. New York and London: The Guilford Press.

Bielschowsky R (1988) *Pensamento econômico brasileiro. O ciclo ideológico do desenvolvimentismo*. Rio de Janeiro: Contraponto.

Birch K (2017) Rethinking value in the bio-economy: Finance, assetization, and the management of value. *Science, Technology, & Human Values* 42(3): 460–490.

Black F and Scholes M (1973) The pricing of options and corporate liabilities. *Journal of Political Economy* 81(3): 637–654.

Brenner N (2000) The urban question as a scale question: Reflections on Henri Lefebvre, urban theory and the politics of scale. *International Journal of Urban and Regional Research* 24(2): 361–378.

Brenner N (2004) *New state spaces: Urban governance and the rescaling of statehood*. Oxford: Oxford University Press.

Brenner N and Theodore N (2002) *Spaces of neoliberalism: Urban restructuring in Western Europe and North America*. Oxford: Blackwell.

Brenner N; Peck J and Theodore N (2010) Variegated neoliberalization: Geographies, modalities and pathways. *Global Networks* 10(2): 182–222.

Brown JC and Purcell M (2004) There is nothing inherent about scale: Political ecology, the local trap and the politics of development in the Brazilian Amazon. *Geoforum* 36: 607–624.

Castells M (1977) *The urban question*. London: Edward Arnold.

Christophers B (2018) Risking value theory in the political economy of finance and nature. *Progress in Human Geography* 42(3): 330–349.

Costa FN da (2012) *Brasil dos Bancos*. São Paulo: Editora da Universidade de São Paulo.

Dymski GA (1998) "Economia de bolha" e crise financeira no Leste Asiático e na California: uma perpectiva espacializada de Minsky. *Economia e Sociedade*, Campinas 11: 73–116.

Dymski GA (2009) Afterword: Mortgage markets and the urban problematic in the global transition. *International Journal of Urban and Regional Research* 33(2): 427–442.

Faulhaber GR and Baumol WJ (1988) Economists as innovators: Practical products of theoretical research. *Journal of Economic Literature* 26(2): 577–600.

Fiori JL (1997) *Os moedeiros falsos*. Petropolis: Vozes.

Greco E (2015) *Value or rent? A discussion of the research protocol from a political economy perspective*. LCSV Working Paper Series No. 8. Leverhulme Centre for the Study of Value, the University of Manchester, Manchester.

Harvey D (1973) *Social justice and the city*. Baltimore: John Hopkins University.

Harvey D (1982) *The limits to capital*. Oxford: Basil Blackwell and Chicago: University of Chicago Press.

Harvey D (2014) *Cidades Rebeldes. Do Direito à Cidade à Revolução Urbana*. São Paulo: Martins Fontes.

Helleiner E (1994) *States and the re-emergence of global finance: From Bretton Woods to the 1990s*. Ithaca and London: Cornell University Press.

Hilferding R (1910) *Finance capital*. London: Routledge & Kegan Paul (1981 Edition).

Jessop B (2000) The crisis of the national spatio-temporal fix and the tendential ecological dominance of globalizing capitalism. *International Journal of Urban and Regional Research* 24(2): 323–360.

Jessop B (2015) Hard cash, easy credit, fictitious capital: Critical reflections on money as a fetishized social relation. *Finance and Society* 1(1): 20–37.

Klink J (2014) The hollowing out of Brazilian metropolitan governance as we know it: Restructuring and rescaling the developmental state in metropolitan space. *Antipode* 46(3): 629–649.

Klink J and Denaldi R (2014) On financialization and state spatial fixes in Brazil: A geographical and historical interpretation of the housing program My House My Life. *Habitat International* 44: 220–226.

Lapavitsas C (2013) *Profiting without producing: How finance exploits all of us*. London and New York: Verso.

Law-Yone H (2007) Another planning theory? Rewriting the meta-narrative. *Planning Theory* 6(3): 315–326.

Lefebvre H (1974) *La production de l'espace*. Paris: Anthropos.

Maricato E (1996) *Metrópole na Periferia do Capitalismo: Ilegalidade, Desigualdade e Violência*. São Paulo: Hucitec.

Markowitz HM (1959) *Portfolio selection: Efficient diversification of investments*. New Haven: Yale University Press.

Marx K (1967) *Capital*. Vol. 1, Third Edition. London: Lawrence and Wishart (1862–1863).

Massey D (2007) *World city*. Cambridge: Polity Press.

Moulaert F (2005) Institutional economics and planning theory: A partnership between ostriches? *Planning Theory* 4(1): 21–31.

Nitzan J and Bichler S (2009) *Capital as power. A study of order and creorder*. London and New York: Routledge.

North D (1990) *Institutions, institutional change and economic performance*. Cambridge: Cambridge University Press.

Oliveira F de (1988) O surgimento do Antivalor. Capital, força de trabalho e fundo público. *Novos Estudos* 22: 8–28.

Paulani LM (2011) A inserção da economia brasileira no cenário mundial: uma reflexão sobre o papel do Estado e sobre a situação atual real à luz da história. In: *Logros e Retos del Brasil Contemporâneo*. Cidade de México, México, 24 a 26 de Agosto de 2011. UNAM.

Peck J (2017a) Transatlantic city, part 1: Conjunctural urbanism. *Urban Studies* 54(1): 4–30.

Peck J (2017b) Transatlantic city, part 2: Late entrepreneurialism. *Urban Studies* 54(2): 327–363.

Peck J (2018) Teoria Urbana e Geografia Econômica: para além da comparação? In: Brandão CA; Fernández VR and Ribeiro LC de Queiroz (eds.) *Escalas espaciais, reescalonamentos e estatalidades: lições e desafios para América Latina*. Rio de Janeiro: Letra Capital: Observatório das Metrópoles, pp. 167–222.

Peck J and Whiteside H (2016) Financializing Detroit. *Economic Geography* 92(3): 235–268.

Pryke M and Allen J (2017) Financialising urban water infrastructure: Extracting local value, distributing value globally. *Urban Studies* 56(7): 1326–1346.

Purcell TF; Loftus A and March H (2019) Value-rent-finance. *Progress in Human Geography*. Online First: 1–20. Available at: https://journals.sagepub.com/doi/10.1177/0309 132519838064 (Accessed 5 May 2019).

Robinson J (2011a) Cities in a world of cities: The comparative gesture. *International Journal of Urban and Regional Research* 35(1): 1–23.

Robinson J (2011b) The travels of urban neoliberalism: Taking stock of the internationalization of urban theory. *Urban Geography* 32(8): 1087–1109.

Royer L (2014) *Financeirização da Política Habitacional: Limites e Perspectivas*. 1st Edition. São Paulo: Annablume.

Savini F and Aalbers MB (2016) The de-contextualization of land use planning through financialization: Urban redevelopment in Milan. *European Urban and Regional Studies* 23(4): 878–894.

Sanfelici D and Halbert L (2018) Financial market actors as urban policy makers: The case of real estate investment trusts in Brazil. *Urban Geography*. Online First. Available at: www.tandfonline.com/doi/full/10.1080/02723638.2018.1500246 (Accessed 5 May 2019).

Soja EW (1989) *Postmodern geographies: The reassertion of space in critical social theory*. London: Verso.

Stiglitz JE (2007) *Making globalization work*. New York and London: WW Norton and Company.

Swyngedouw E (1997) Neither global nor local: "Glocalization" and the politics of scale. In: Cox KR (ed.) *Spaces of globalization: Reasserting the power of the local*. New York and London: The Guilford Press, pp. 115–136.

Tavares M de C (1999) Império, Território e dinheiro. In: Fior JL (ed.) *Estados e Moedas no desenvolvimento das nações*. Petrópolis, RJ: Vozes, pp. 449–489.

Theurillat T and Vera-Büchel N (2016) Commentary: From capital lending to urban anchoring: The negotiated city. *Urban Studies* 53(7): 1509–1518.

Torrance MI (2009) The rise of a global infrastructure market through relational investing. *Economic Geography* 85(1): 75–97.

Villaça F (1999) Uma contribuição para a história do planejamento urbano no Brasil. In: Deák C and Schiffer SR (eds.) *O processo de urbanização no Brasil*. São Paulo: Editora da Universidade de São Paulo, pp. 169–244.

3 State spaces, urban rents and finance in the Brazilian development trajectory

3.1 Introduction

Table 3.1 provides a stylized representation of the financial-monetary geo-history (Soja, 1989) of the Brazilian development trajectory from the early nineteenth century until recently. Row 1 summarizes the imbricated relations between the state and the economic-spatial transformations that have occurred over time. In line with the previous theoretical-conceptual discussions on financialization, for the specific purpose of our analysis we have added a dimension on monetary and financial governance (second row) and another one on the role of the state in the finance of cities (third row). Our objective is to trace more clearly the connections and interdependencies, if any, between the nationally scaled project of industrialization, import substitution and urbanization, and the financial-monetary mechanism through which this was realized. The latter includes an analysis of the socio-spatially selective and contradictory funding mechanism that were mobilized in the building, operation and maintenance of housing and infrastructure networks as essential conditions for the success of the national developmental project as such. This is also relevant in understanding the differences between the earlier discussed role of the public fund under spatial Keynesianism in the Global North (i.e. representing something of an anti-value, meaning that the assumptions that drive the reproduction of value suggest a denial of the system's fundamentals) as compared to its supposed equivalent during the Brazilian national development regime.

While the objective here is not to present a detailed review of the historical and institutional literature, Table 3.1 nevertheless provides a particular analytical reading of the financial geo-history of the Brazilian development trajectory that allows a broader understanding of the entanglements between the state, money and credit in the production of urban space. Moreover, the table serves the purpose of filling in some of the gaps in the existing critical literature in between strands influenced by political economics (that lack more detailed discussions on spatial-urban dimensions) and urban studies (usually having a more distant look on issues related to credit and finance).

Table 3.1 Brazilian spatial development trajectory and financial-monetary regimes

Period	1800s–1915/30	1915/30–1964	1964–mid 1980s	mid 1980s–2003	2003–2015
The State and the country's geo-historical development trajectory (Section 3.2)	Rural agro-export oriented regime (resource-intensive and unprocessed products, (e.g. sugar, rubber, coffee); Smaller urban centres perform the role of archipelagos-territorial platforms for exports; Emerging role of newly industrializing region of the state of São Paulo in national space economy.	Emerging metropolitanization & rise of national developmentalism: industrialization and import substitution (non-durable consumer goods); Polarization of the national space economy by the State of São Paulo.	Intensification of metropolitanization; Authoritarian national developmentalism; Design and implementation of comprehensive development plans (at national, regional and metropolitan-urban scale) aimed at the consolidation of a national space economy and the complementary stages of industrialization (consumer durables; intermediate and capital goods).	Neoliberalization, democratization, restructuring, rescaling and roll-back of national developmental regime; Modernization and downsizing of domestic industry; internationalization of production chains (e.g. car manufacturing); spatial fragmentation and competitive regimes (tax wars); competitive regulatory downgrading; hollowing out of metropolitan governance and shared institutional arrangements and so on.	2003–2013: Social/ neo-developmental regime; re-emergence of national scaled policies for industrial and technological development; National framework on urban policies (Ministry for Cities); 2014–2015: economic and political crisis and presidential impeachment.

(Continued)

Table 3.1 (Continued)

Period	1800s–1915/30	1915/30–1964	1964–mid 1980s	mid 1980s–2003	2003–2015
Monetary and Financial Governance (Section 3.3)	Absence of central bank and consolidated domestic institutions for money, credit and finance; monetary circulation based on commodity money (slaves, coffee and gold) linked to international trade and FDI; 1850: prohibition of slave traffic; 1851: land reform; predominance of international banks in speculative IPO and founder's profit, funding loans (coffee), government debt and urban infrastructure finance.	1930–45: Emergence of national banking network under flexible capital requirement regulations; stricter regulation of international banks and protection of national financial system; creation of the first public banks (state and federal); 1945: *Banco do Brasil* and *Superintendency on Money and Credit (SUMOC)* assume the coordination of the monetary and the credit system; 1945–1964: initial concentration of banks; incipient national regulatory framework on deposit/reserves.	1964: Creation of Central Bank and monetary reform; Indexation of contracts and creation of national system for housing and infrastructure finance (HFS); centralized supervision of monetary system (deposit-reserves; open market operations and so on; "monetarist" influences on domestic policies); 1964–1974: Failed attempts to introduce US-inspired stock exchange; 1974–1988: crisis, restructuring and concentration of national banking.	Moratorium, Latin American debt crisis, economic stagnation and inflation (1980s); stabilization of inflation through Plano Real (1994); liberalization of domestic capital markets and implementation of Incentive Program aimed at the Restructuring and Strengthening of the National Financial System (*PROER*) and privatization of state banks; federal downloading of austerity programs (Chapter 7); concentration of domestic banking; gradual re-emergence of internationalization of banking system (crisis-driven procurement of cheap assets).	Strengthening of the role of federal banks in industrial, technological and urban development policies (Banco do Brasil; BNDES and CEF); Financial inclusion of low-income groups (*bancarização*); Market-based social housing provision through the My House, My Life (*PMCMV*) program; subsidies for slum upgrading (National Growth Acceleration Program (PAC)); Dilma Presidency: policies aimed at lowering interest rates & proliferation of tax incentives for national industry; continuing internationalization of banking driven by the perception of opportunities in a growing Brazilian market.

(Continued)

Table 3.1 (Continued)

Period	1800s–1915/30	1915/30–1964	1964–mid 1980s	mid 1980s–2003	2003–2015
State spaces, the public fund and finance of cities (Section 3.4)	Absence of state finance for housing and urban development; private rental housing regime and industrial workers housing estates; international private finance for investment and O&M of infrastructure networks (energy and logistics, urban transportation, water supply).	Collaborative arrangements between state and local governments in infrastructure (e.g. water and basic sanitation): state and municipalities share ownership of metropolitan-specific purpose public utilities–see Chapter 5); Decentralized, small-scale cooperative housing schemes (Institutes for pension and retirement funds– IAPs); Failed national state initiatives for housing finance (e.g. National Foundation for Low-Income Housing–*Fundação Casa Popular*–1946).	Consolidation of national housing finance system based on compulsory worker's contributions (FGTS) and voluntary savings and loans (SBPE); PLANASA (National Water Supply and Sanitation Plan) and National Housing Bank (NHB); Separation between Social Housing Finance System (funding through FGTS) and Market Housing Finance System (funding through SBPE–modelled on the US Savings and Loans Associations).	Bankruptcy of the NHB and demise of PLANASA; Institutional vacuum regarding finance and governance of urban-metropolitan areas; After *Plano Real*: Discussion and design of market enabling legislation; 1997: SFI–Real Estate Finance System; Pressure to privatize-deregulate public infrastructure companies (e.g. SABESP–see Chapter 5).	Institutional strengthening of housing and urban development (City Statute; National System of Social Housing; National framework law on basic sanitation; See Chapter 5; Significant inflow of budgetary resources and subsidized credit for housing and infrastructure through PMCMV (market-built social housing–see Chapter 6) and PAC (slum upgrading).

Source: Author's elaboration. Based on Costa (2012), Klink (2013) and Klink and Keivani (2013).

3.2 The state and the country's geo-historical development trajectory

Table 3.1 can be read column or row-wise. In this section we will discuss the first row related to the country's geo-historical development trajectory. The next two sections deal with monetary-financial governance and the role of the state and the public fund in the finance of cities, respectively.

The first row summarizes the country's transition from a colonial, export-led agricultural system[1] into a capitalist national space economy. This involved successive stages of industrialization, import substitution and what Lefebvre (Brenner and Elden, 2009: 223) called a "state mode of space production" through the planning and implementation of a national network of complementary regions and cities. In the particular Brazilian setting, the latter implied the gradual trans-formation of isolated regional economies, as export-oriented archipelagos, into a national space economy, which was located and grounded regionally through a network of complementary production circuits (Araújo, 1999; Oliveira, 1990). To be clear, regional clusters had been abundant throughout the Brazilian colo-nial geo-history. They were characterized by recurrent, short-lived and predatory cycles (from the point of view of the rapid depletion of natural resources and the deterioration of eco-systems and living conditions), structured around products such as gold and iron ore, sugar, rubber and coffee, among others (Prado, 1994). For the first time, however, the national developmental project constituted an incipient series of complementarities among Brazil's regional economies.

Particularly during the military regime (1964–1985), a significant effort, in terms of the planning and financial resources, was put into metropolitan areas. The latter became the privileged platforms of a top-down institutional and finan-cial framework that would gradually transform metropolitan areas into the spatial backbone and nerve centre of the national developmental strategy of industrial-ization and urbanization. One of the obvious side effects of this developmental effort was an increase in socio-economic disparities among the states. Despite the efforts that were launched in order to reduce "the backwash effects" (Myrdal, 1972) of growing macro-regional disparities, which were particularly focussed on the poorer and less dynamic states of the Northeast, the national industrial space economy that emerged during this period was nevertheless clearly under the hierarchical command of the state of São Paulo in general, and metropolitan São Paulo in particular (Cano, 1998). During 1930–1970, despite the nationally structured compensating policies to the poorer regions, the state of São Paulo clearly emerged as a net importer of financial and human resources from the rest of the country (Oliveira, 1981).

From the mid-1980s onwards, decentralization, neoliberalization and recurrent economic crises triggered successive and contradictory stages of rescaling, roll-ing back and out of the developmental state. This process both mobilized and was influenced by cities and metropolitan areas as increasingly central spaces for the production and appropriation of value and reproduction of life itself. In the remainder of this chapter we will analyze with more historical detail the entangle-ments between the state, money and finance in the (re)production of space.

3.3 Monetary and financial governance

The second row of Table 3.1 is about monetary and financial governance. Under "the agricultural export-led regime (1800s-1915/1930)", the country's domestic monetary system was limited and mainly served as a support to the coordination of trade flows, the exchange rate and the profit remittances from its dynamic export regions to the core countries of the global economy.[2] International banks centralized this articulation, including bringing in their own foreign staff, while minimizing overseas investments (Costa, 2012). The absence of an integrated domestic economy with a complementary network of cities also implied a weakly developed local system of fiduciary and bank money.

It is nevertheless worthwhile to analyze in more detail why, during this period, Brazil didn't manage to create a domestic system of money, credit and finance on the basis of the surpluses that were being generated by its export-led agricultural economy.[3] A first, more obvious response is that, during a large part of the nineteenth century, as a Portuguese colony, Brazil was not an independent nation state and, as such, was in no condition to control the emission of its domestic money supply.[4] Consequently, the centralization of monetary policy by the Portuguese was more aligned with the conspicuous consumption of the court and the drainage of Brazilian wealth, as opposed to the constitution of a modern system to finance the production, commercialization and exportation of agricultural products such as coffee.

But this "centre-colonial periphery" biased argument only partially explains a more complex picture. Particularly from the second part of the nineteenth century until World War One, Brazil lost a series of opportunities, more specifically during the initial stages of its transition from a rural slavery and mercantile system, based on coffee, into an urban-industrial and capitalist economy. In that respect, it is useful to revisit some of its history. In 1850, the international traffic of slaves was prohibited, while Brazil was only to abandon slavery in 1889. Not coincidently, one year after the prohibition of the traffic, a land reform was enacted in 1851. It formalized the clientelist-patrimonial land possessions from the colonial system and at the same time constituted the legal-institutional mechanisms of land markets, as well as the possibility to use land as collateral in credit operations. In practical terms, the land reform generated rents to the formally proclaimed property owners, while raising the barriers for ex-slaves and foreign immigrants to access land (Martins, 2010). Moreover, the abolition of the international traffic of slaves generated both a wealth effect for existing slave-owners (considering the increased scarcity of slaves and the expectation of associated price increases) and a short-run excess supply of loanable funds (released from the international traffic) awaiting alternative profitable investment options. Fundação Perseu Abramo (FPA, 2019) analyzes how, in the particular context of the state and metropolitan economy of São Paulo, the emerging capitalist coffee cluster could have provided, at least theoretically, opportunities for the design and rollout of domestic arrangements for the finance of production, commercialization and exportation of its products. In a subsequent stage, this could potentially have enabled a move from short-term supplier's credit to long-term finance (through the stock markets

88 State spaces, rents and finance in Brazil

and bank credit) aimed at the expansion of the coffee cluster, infrastructure invest-ments and, eventually, industrialization itself.[5]

Nevertheless, this never happened. Although there was some involvement of local commissioners, clearing houses and other financial intermediaries in the circulation of credit and fiduciary money aimed at the leverage of the coffee cluster in metropolitan São Paulo and the western part of the state of São Paulo, these efforts were not up-scaled. Instead, the bulk of the articulation between the provision of short-term supplier's credit and the finance of international trade was maintained under the hierarchical command of international banks. As a matter of fact, the owners of the semi-capitalist coffee plantations were more interested in the accumulation of wealth and generation of rents, a strategy that had become viable through the emerging nexus between the markets for land, slaves and coffee. Thus, the industrialization of the country's dynamic regions (metropolitan São Paulo and the interior of the state) was not accompanied by the gradual constitution of domestic markets for shares and credit; in a way, then, industrialization eventually occurred "not on the basis of local banking, but in spite of it" (FPA, 2019: 78). At the same time, the allocation of domestic loanable funds through strategies oriented towards obtaining urban rents and founder's profit represented something of an "early urbanization of finance" (along the lines of the conceptual discussion of the previous chapter), that is, occurring before the consolidation of a full-fledged network of cities and met-ropolitan areas.

A final point in understanding the limited penetration of domestic institutions for credit and finance during this period is that even after its formal "indepen-dence" from Portugal (in the year of 1889), the Brazilian state was effectively committed to supporting the role of international finance within the prevailing global monetary order, which was based on maintaining a fixed parity between the Brazilian currency and the price of gold (according to the gold standard).[6] This implied a continuous inflow of international resources in order to finance govern-ment debts (concentrated in the capital city of Rio de Janeiro) and short-term sup-pliers' credit and funding loans (concentrated in the state and metropolitan area of São Paulo), frequently backed up by generous interest payments indexed to the gold standard (Costa, 2012: 351). Particularly the English investors operating in the "city" evaluated the Brazilian option an attractive one in terms of the low risks, predictability and (exchange rate) stability for the repatriation of foreign dividends and profits.

Thus, Topik (1979: 404) summarizes the situation in 1912 in the following terms:

> individuals, and not credit institutions, guaranteed 82,5% of the value of all outstanding loans. These initiatives resulted, most likely, from activities articulated by coffee merchants and financial intermediaries aimed at the supply of short-term trade-credit, instead of long-term finance. The banking system of the republic, particularly in the beginning, was neither adequate to stimulate industrialization, nor to finance even a relatively sophisticated and prosperous agriculture sector.

Two historical episodes serve to illustrate how these missed opportunities in terms of constituting consolidated domestic markets for credit and finance were both cause and consequence of the dominant role of international banking during the country's export-led development trajectory. The first is related to the management of recurrent crises in its export-oriented coffee economy (Furtado, 2005). During the period from the late nineteenth until the early twentieth century, under the prevailing gold standard, Brazil was regularly facing crises in the stabilization of its domestic exchange rate and the export prices of coffee. The "solution" that was found consisted in mobilizing the state around the stabilization of coffee prices through the procurement of excess stocks. While this represented "pre-Keynesian" style income and employment generation, it also provided coffee producers a differential rent in light of the possibility to accommodate production levels and speculate on the likely timing and degree of future price increases (Miranda and Tavares, 1999; Tavares, 1999). Moreover, in order to avoid an increase in non-convertible domestic money and subsequent exchange rate devaluations, which would be incompatible with the gold standard, the state realized its procurement operations through successive loans that were intermediated by the English banking system. In a way then, foreign banks were actively involved in the generation of (domestic) suppliers' rents and interest payments on the outstanding funding loans that were contracted in order to finance the stabilization of coffee prices.

A second episode refers to the proliferation of Initial Public Openings of existing Brazilian companies that were being listed on the European and London stock exchanges. The process was coordinated and intermediated by foreign banks and represented a clear example of what Hilferding had described as founder's or promoter's profit (Hilferding, 1910; Costa, 2012). In other words, instead of mediating additional foreign direct investments in Brazil, foreign banks leveraged their financial profits through the exploration of the difference between the relatively small volume of funds that were required to set up a company as compared to the relatively higher present value of dividends that could be obtained by floating these companies at the European stock exchange.[7]

The emergence of the "national development regime (1915/30–1964)" represented a clear break away from the previous regime of monetary governance, which had created an enabling environment aimed at the "internationalization of the Brazilian economy" through the "internalization of foreign finance and banking" under the gold standard (Costa, 2012). Initially this meant stricter regulation of international banks regarding the required hiring of domestic staff as well as the mobilization of foreign investments in setting up local branches.[8] This was subsequently followed by domestic market making (e.g. lowering regulatory barriers for setting up domestic banks) and protection, culminating in the nationalization of new foreign banks in the 1937 constitution, while existing international financial institutions were allowed to continue their operations under stricter guidelines (Costa, 2012: 381).[9] Although it would last until 1964 before the country established its central bank, the challenges involved in building up an industrial complex through import substitution in a context of continuous trade deficits required strengthening the domestic control over monetary, trade and exchange rate policies. This became the joint responsibility of the state bank *Banco do Brasil* and the super-intendancy for money and credit (Rangel, 1981). The federal government also created its first public banks aimed at targeting specific sectors such as energy, steel and petro-chemical industries and providing financial support to its national developmental project. For instance, the National Bank for Economic Development (BNDE) was created in 1952.[10]

The "military regime (1964-mid 1980s)" designed and consolidated a full-fledged reform of the monetary and financial system. This included the creation of a central bank that was modelled along the lines of US monetarist thinking on the role of this institution, equipped to implement open market operations and supervise credit-deposit ratios aimed at the control and supervision of the money supply. Considering the rising levels of inflation, the government also designed a system of indexation that would guarantee the financial viability of long-term credit in sectors with long payback periods such as housing and urban infrastructure. Nevertheless, as will be discussed subsequently, the indexation of long-term contracts would only prove a second-best solution for the continuous erosion of the value of domestic money.[11] A final element worth mentioning in this period was the first attempt to constitute a stock market along the lines of the US experience. This would prove largely unsuccessful considering the prevailing financing pattern of national industry through both family groups, which by and large relied on reinvested profits, as well as state support (Costa, 2012; Tavares, 1999). In the meantime, accelerating inflation, macroeconomic stagnation, the international debt crisis of the 1980s and the bankruptcy of institutions such as the National Housing Bank (in 1985) set the stage for an intense, crisis-driven restructuring of financial and monetary governance which would take place during much of the 1990s.

The 1994 Plano Real eventually proved instrumental in eliminating inflation through the liberalization of international trade and capital markets and the creation of a new currency (the Real) after a short phase of transition. Influenced by the international circulation of ideas regarding the role of currency boards in emerging countries' macroeconomic stabilization,[12] it was based on an (overvalued) exchange rate pegged to the dollar, liberalization of foreign trade and capital accounts and a fast rollback of "old" developmental industrial and technological policies. These mechanisms triggered lower import prices, which, in combination with the liberalization of the trade and capital account, would be essential in maintaining control over the escalation of domestic price levels. Contrary to initial expectations that it would last a couple of months, the Plano Real proved surprisingly robust but came at a cost of generating extremely high domestic interest rates in order to guarantee the continuous inflow of international reserves.

History doesn't repeat itself. Nevertheless, and even acknowledging that the monetary and banking system of the country under the gold standard until the First World War was rudimentary, during the last three decades or so (mid 1980s-mid 2000s and mid 2000s-2015), Brazil has seen a re-emergence of the role of global finance and international banks in the articulation of its space economy with the rest of the world (Costa, 2012).

During the 1990s this was based on a crisis-driven entrance of international banks and global finance. These players envisaged financial gains in light of the national banking crisis and reform, which triggered the outright sale of public banks, privatization of infrastructure assets in sectors such as energy and tele-communications and a depreciating value of domestic assets associated with a stagnating economy.

From the mid-2000s onwards, however, internationalization continued for reasons that were related to international banking's perceptions regarding the growing market opportunities driven by the social developmental regime itself. These opportunities arose in light of the increasing rollout of policies toward financial inclusion, the growth of public subsidies for infrastructure (particularly energy, logistics and urban infrastructure required for slum upgrading) and housing, as well as the country's improved international credit rating. Thus, international players such as Santander, ABN-AMRO, Citi Bank and HSBC, among some of the well-known examples, have significantly increased their activities in the Brazilian market since then (Costa, 2012).

3.4 State spaces, the public fund and finance of cities[13]

The third row of Table 3.1 provides a schematic reading of the entanglements between the developmental state restructuring and rescaling and urban finance in Brazil.

As mentioned, under the "agricultural export-led regime (1800 and 1915–1930)", Brazil was a rural country with a few emerging urban centres that would concentrate its efforts towards industrialization in a subsequent stage. During this pre-industrial era, the structure of housing provision and finance was directly related with the earlier-mentioned transitions that were occurring in the country. To be specific, the constitution of land markets and the possibility to use land assets as collateral, the prohibition of the traffic of slaves and the consolidation of the regional coffee complex in São Paulo generated a scenario whereby available loanable funds could be allocated and invested in rental housing in order to support the growing urban centres. Rental and precarious tenement housing represented prime investment options for domestic capital and were responsible for the bulk of the housing provision. This was complemented with relatively small-scale housing estates that were provided (and charged) by industrialists to their workers. Owner-occupied housing was still marginal.

At the same time, international capital performed an important role in the private finance, provision and operation of profitable urban infrastructure such as transportation, basic sanitation, energy and lighting and railways, among other examples. The state was by and large absent in urban planning and was only involved in the monitoring of sanitary conditions through inspection and/or demolition of substandard housing.

The "initial stage of the national developmental model (1930–1964)" was marked by a populist pact between the emerging urban-industrial bourgeoisie and the labour movement, structured around the design and implementation of a developmental project of industrialization and urbanization (Bielschowsky, 1988). Nevertheless, this strategy was soon to face its dilemma of how to provide relatively expensive workers' housing without increasing production cost and crowding out investments that were badly needed to support the growth of a national infant industry. Therefore, industrialists and policy makers agreed that housing was a strategic sector, both in terms of its capacity to generate income and employment and to mobilize political support for the national-developmental project. As such, it couldn't be left to the market but required active state organization and intervention (Bielschowsky, 1988; Bonduki, 1998). This was easier said than done, however. It would take several decades of smaller-scale initiatives before a first national housing system was established in the mid-1960s.

The Institutes for pension and retirement funds (i.e. the *Institutos de Aposentadoria e Pensões*, ISPs), which were constituted in the 1940s in sectors such as banking, commerce, industry, shipbuilding and energy, represented the first of these smaller-scale state-mediated initiatives.[14] Their funding was based on a surcharge on wages, which was channelled into the financing of both rental and owner-occupied housing to members or external institutions.[15] Although the architectural quality of some of its housing estates was undeniable, the effective outreach of the Institutes was relatively small considering the lack of scale. Moreover, their financial arrangement, structured around retirement and pension schemes, was highly selective: while informal workers were excluded altogether, the institute's financing guidelines never broke with the premise of cost-recovery, thereby effectively moving away from the officially proclaimed objective of providing social housing to its members. In the meantime, the gradually rising inflation rates led to credit rationing, which implied an increasingly privileged access to funds for contributing and employed members of the Institutes.

The National Foundation for Low Income Housing (that is, the *Fundação Nacional Casa Popular*) was created in 1946 and represented another, rather unsuccessful effort to design and implement a full-fledged national institutional framework for housing and urban development finance. By many it was considered to be a laboratory of what would become the National Housing Bank in the 1960s. Its ambition was to move beyond the autonomous, cost-recovering and corporatist approach of the six Institutes for pension and retirement funds. In order to do so, it would centralize the institutes and create a built-in mechanism for cross-subsidization, based on a 1% surcharge on all real estate transactions.

Rather unsurprising, the six Institutes for pension and retirement funds were opposed to the creation of the foundation. As a matter of fact, the prevailing decentralized arrangement suited them quite well considering they were operating on an autonomous and profitable basis. Real estate developers also contested the formula for cross-subsidization, considering it as a threat to their profit margins. The failure of the effective take-off of the National Foundation for Low Income Housing represented something of a delay of twenty years in terms of housing policy and finance (Bonduki, 1998). In the meantime, with the Brazilian urban transition in full swing, the national rent control legislation (1942) represented a landmark in the transformation from rental to owner-occupied housing, considering this measure both reduced profitability and drained resources and investments from the rental sector.

The coup of 1964 would be followed by two decades of military regime (1964–1985). It was triggered by intense internal class pressures within the earlier-mentioned populist national developmental pact, which led to demands for more radical and structural reforms, such as the distribution of land in rural and urban areas. This was aggravated by a rapidly deteriorating economic crisis that undermined the viability of the populist Goulart administration.

Pressured by escalating housing costs, a rapid growth of urban centres and slums and a shrinking rental sector, one of the first flagship projects of the military regime during 1964–1967 was the financial-institutional design of a national system for housing and urban development. It was structured around a centralized National Housing Bank (NHB) and associated institutions for credit. The immediate objective was to roll out in a more ambitious way the experimentation that had taken place during 1930–1945 with owner-occupied housing finance through the retirement and pension schemes. The strategy was not only aimed at increasing the leverage of the national developmental project through the strengthening of a domestic market for building, construction, civil engineering and infrastructure services, triggering positive multiplier effects on income and employment generation. It also envisaged disseminating an ideology of owner-occupied housing as an important element for the political stabilization of the country and legitimization of the regime. This was also instrumental in reinforcing the country's recently acquired status of a rapidly urbanizing newly industrializing country, which was going through a growth miracle and required additional international financial support.

As mentioned, the first step was to design methods that would guarantee the indexation of outstanding loan balances and applications, which had proven to be one of the vulnerabilities of the previous system. Moreover, the new arrangement introduced what would become a long-standing feature of Brazilian housing finance, based on two parallel circuits of market-based housing, which was funded by voluntary savings and deposits, and a social housing sector that received resources from compulsory worker's contributions on the basis of a surcharge on wages during their working life. The latter was the so-called *Fundo de Garantia de Tempo de Serviço* (FGTS, meaning the Guarantee Fund Based on Service Time Contributions) and would become the main source of social housing finance. It is important to stress that, different from the role of the public fund under spatial Keynesianism in the context of the Global North, the FGTS didn't receive subsidized budgetary state funds and was based on the recovery of (albeit partially subsidized) funds "mobilized by workers themselves" (Eloy et al., 2013).

Initially, the Housing Bank also experimented with the creation of a national mortgage program for social housing. At the end of the construction period, originating banks would pass mortgages and the associated risk of loan defaults to the National Housing Bank. Despite these efforts, it was never able to trigger a deep and liquid market.

There is by now abundant literature that has critically evaluated the significance of the experience of the Housing Bank during the developmental regime (Bolaffi,

1977; Maricato, 2011; Mello, 1988). The bank never managed to provide afford-able and well-located housing finance to low-income groups on a sustainable and ongoing basis. Its approach of delegating the allocation of finance, as well as the design, construction and location of housing units to banks, mortgage originators and real-estate developers failed considering the enormous loan defaults, mismatches between demand and supply and a mass production of poor-quality housing located in isolated places in the outskirts of metropolitan areas, without adequate infrastructure. Moreover, considering that its main source of funding was based on cost recovery, the bank effectively shifted its portfolio to medium-income borrowers and, particularly after the debacle of the Mortgage Program, to investments in financially more viable urban infrastructure such as transportation, water and sewage.[16] Finally, the Housing Bank performance became something of a paradigm in terms of the mismatch between housing finance and urban policy. It had generated huge isolated housing estates in the outskirts of city regions without access to urban services and opportunities for income and employment generation. As will be discussed in Chapter 6, this dramatic legacy of the NHB is now being revisited, for good and bad reasons, in light of the critical reviews of the more recent My House, My Life program (launched in 2009).

From the mid-1980s onwards, the economic and political demise of the Brazilian national developmental state was accompanied by escalating inflation rates, rising fiscal and current account deficits and intense calls for re-democratization. This evidently also shook up the national housing system. The 1985 bankruptcy and extinction of the National Housing Bank created a vacuum in terms of national policy initiatives and financial resources, generating institutional destabilization and ad-hoc decentralization of responsibilities to local bodies (Arretche, 1996; Valença and Bonates, 2010). This scenario would prevail until the mid-1990s when monetary stabilization through the Plano Real was finally established in 1994.

As mentioned, overvalued currencies such as the real and high domestic interest rates were two sides of the same coin; however, nominal yearly rates were never less than 20% during the 1990s and in 1997 surpassed 40%. This raised critical debates on de-industrialization and the increasingly financialized pattern of the Brazilian economy as collateral damages of the Plano Real (Paulani, 2008). Housing finance in general, and for low-income groups in particular, completely dried up.

At the same time, however, low and stable inflation had reduced investor's risks and opened up perspectives for the leverage of long-term private credit markets. This had also put the country back onto the radar of mainstream international housing and urban development finance (World Bank Institute, 2003). However, the legacy of the NHB in terms of legal insecurity, particularly regarding the absolute size of outstanding debts of existing contracts that had been accumulating since 1985, threatened the building up of "a clean sheet". To be clear, since the beginning of the 1980s the NHB had been working with a sophisticated double-indexed system whereby repayments increased according to salaries, while outstanding loan balances accompanied effective inflation.[17] The system's design was underpinned by the belief of international creditors-donors and the government that macroeconomic structural adjustment would ultimately "deliver" up its promises: the cost of initially shrinking real salaries would be followed up by real growth in monthly incomes (and repayments of contracts), eventually leaving manageable balances at the end of the loan's term. As known, Brazil's failed structural adjustment implied huge debts that were way beyond the capacity of the insurance fund FCVS[18] that had been designed for "moderate" yearly inflation rates of up to 40% when setting up the NHB (Granja, 2008). In the 1990s, the fund's uncontrolled and escalating stock of debt was increasingly perceived as a time bomb that would also hinder the buildup of liquid and deep long-term credit markets. After years of uncertainties, the "solution" found in 1996 was to set up a procedure, formalized by law in 2000, according to which banks and creditor institutions could claim their credits against the FCVS fund. Claims would then be evaluated and accounted for, restructured and gradually transformed into government-backed securities with a maturity of 30 years and paying fixed interest rates.[19] According to estimates from the Ministry of Finance, which is formally responsible for the restructuring and securitization of old NHB debts, between 1998 and 2015, some R$ 127 billion worth of papers have been emitted (Ministry of Finance, 2015). Public stakeholders such as *Caixa Econômica Federal*, the successor institution of the NHB, hold the majority of these securities.

In a way, this represented a classic case whereby "credit was saved by money" (Aglietta, 2018), considering that highly illiquid and risky loans and, in some cases, operations that had been completely written off by private banks, were bought up and "federalized" by the state, securitized and recycled back "onto the books" within the emerging market-based system. To illustrate, sub-prime federalized FCVS securities represented some two-thirds of the assets and guarantees that were used as credit money, during 1994–1998, in the R$ 20 billion-worth federal program on bank reforms (the so-called Incentive Program aimed at the Restructuring and Strengthening of the National Financial System, (i.e. PROER[20]). The latter was aimed at streamlining and up-scaling the troubled financial sector through the privatization of state banks and closure and restructuring of smaller and inefficient financial institutions, while increasing the scale and solvency of firms with potential to capture the market opportunities associated with low and stable inflation (Granja, 2008). As mentioned, this period also represented something of a momentum for the re-emergence of international finance in the domestic economy, looking for the acquisition of cheap private and public assets in a banking environment that was being "cleaned up" and privatized with active support by the Brazilian state (see also Chapter 7).

The consolidation of the Plano Real was also accompanied by a perception that Brazil "would unlock its improved business environment in order to connect with international tendencies of liberalization and deepening of financial and capital markets in emerging countries" (Klink and Denaldi, 2014: 223). Thus, after the stabilization of inflation, the Cardoso administration was keen to move on to the next stage with a project of regulatory rollout aimed at increasing the participation of the private sector in the delivery of credit for housing and urban development. This momentum was also influenced by the international circulation of ideas regarding the supposed fading out of a cycle of direct state intervention in the allocation of credit, as well as harsh criticism of influential local academic think tanks such as *Fundação Getúlio Vargas* (FGV, 2007) regarding the perceived deficiencies of Brazil's "old", (i.e. state-mediated, developmental housing finance).[21] Their study argued that the Housing Finance System (HFS)[22] was not only highly pro-cyclical – any economic downturn would drain its resources through lower savings and wages and higher withdrawals associated with growing unemployment – but also ineffective through its rigid rules regarding the compulsory earmarked allocations of savings in (low-income) housing and real estate. More specifically, the prevailing regulatory framework required that 65% of all savings deposits would be allocated in housing and real estate (that is, 52% targeted at low-income groups, while 13% of all applications could be invested at market rates).[23]

The 1997 creation of the National Real Estate Finance System (REFS) was aimed at complementing the HFS with a market-based arrangement for housing finance modelled along the lines of the US mortgage system. It created an institutional framework for secondary mortgage markets, securitization, real estate-backed securities and fiduciary alienation, which were designed to provide liquidity, scale and the reduction of risk in an emerging private credit market for real estate and housing. Intense regulatory rollout was instrumental in the creation of this new market. One of the initial moves was to exempt buyers of real estate titles from paying income tax. This was followed by a series of measures that effectively created an increasingly thin line between the "old" HFS system and the emerging circuit of secondary real estate finance promoted by the REFS. While the secondary market was operating on a limited scale, between 2002–2010 the earlier-mentioned guidelines for the compulsory application of 65% of savings deposits from workers into (low-income) housing finance were gradually made more flexible and allowed that mortgage-backed securities, real estate funds and similar products could be accounted for as housing investments, under the condition that they were aligned to the overall objectives of the HFS. As such, in 2011 6.4% of HFS resources were allocated in securities of the REFS. Along the same lines, since 2001, the low-income segment of the HFS, composed of the earlier-mentioned compulsory savings fund from workers (FGTS), was allowed to invest in real estate-backed securities, implying further complementarities between the old state-mediated housing finance system and the new real estate finance complex. In 2011, the authorized budget of the worker's fund FGTS for buying up real estate-backed securities was R$ 2.84 billion (Cagnin, 2012: 28).

In short, part of the resources that were mobilized in order to buy real estate-backed securities didn't effectively represent additional money but a mere reallocation of existing HFS funds (Cagnin, 2012). Moreover, not all of these funds were related to housing in general, or social housing in particular (Royer, 2014), and were channelled into non-residential uses. Finally, in spite of this regulatory rollout, the linkages between residential real estate, housing finance and capital markets were not deepened much during the mid-1990s to the early 2000s. To illustrate, figures for 2012 showed that, despite a more explicit growth pattern of real estate products from 1998 onwards, 93% of the total worth of outstanding housing debt (R$ 195 billion) was still funded by the state-mediated HFS (Cagnin, 2012: 20, 21). High-yielding/low-risk government bonds continued to offer a more favourable and easy opportunity for national and international investors. The innovative financial engineering was also surrounded by doubts from investors regarding the long-term sustainability of inflation control, as well as the mediocre growth of household income and employment. Finally, the system still lacked operating experience.[24]

Nevertheless, while recognizing that the Brazilian real estate finance system of 1997 represented something of a truncated financialization (Shimbo, 2012), its regulatory rollout at least set the stage for a potential take-off for deeper entanglements between finance and real estate, which were effectively stimulated through the application of the compulsory savings funds from the FGTS into the REFS. Particularly considering the growing low-to-medium income housing markets that accompanied macroeconomic growth and the shift in the governmental orientation, this would become relevant in the 2000s. More specifically, from 2003 onward, a renewed macroeconomic growth, income distribution and federal cash-transfer programs to poorer families (such as *Bolsa Familia*, meaning the conditional family cash transfer program linked to schooling), as well as the continued rolling out and institutional strengthening of the state in thematic areas such as technological, industrial and spatial policy, raised a debate on the emergence of what has been called a social-developmental state (Fernandes and Novy, 2010; Klink, 2013). According to this characterization, the state's strategies were marked by hybridity between redistributive-progressive and market-friendly projects of weak social-urban reform and financialization, respectively, which have been filled in at multiple scales and arenas. A brief overview of this contemporary period will serve for the purposes of this chapter, considering that the case studies will delve into the details.

The approval of the earlier-mentioned federal City Statute (2001) was followed by the creation of the Ministry for Cities in 2003. It represented a clear move in the direction of institutional strengthening for urban policy development and implementation. For instance, the Ministry was quick to design and implement a national capacity-building strategy aimed at supporting local governments in the elaboration of their master plans aligned with the premises of the City Statute structured around the provision of well-located and affordable low income housing and the reduction of land speculation. Moreover, the Ministry coordinated the participatory formulation of a national housing strategy, which was based on a national plan, a fund and participatory councils in order to fill in the guidelines of policies. The strategy recognized that increasing the lines of credit to medium-income segments through market-based finance would be instrumental in unlocking a more directly targeted, subsidized and budgetary flow of funds for housing to the lowest income groups (Brasil, 2010). It envisaged that the best way to do that, however, was to insist in leveraging the REFS with funding from the old "developmental" HFS. Thus, in between 2003 and 2008, the federal government continued to stimulate the REFS by building in more flexibility regarding the application of the compulsory savings fund FGTS into market-based housing finance. Consequently, the amount of compulsory savings from workers into the REFS increased from R$ 8 billion in 2005 to more than R$ 40 billion in 2008, a figure which was also reinforced by demand-led macroeconomic growth (Shimbo, 2012).

Three additional state regulatory strategies were also essential to the project of financialization, (i.e. the incorporation of fiduciary alienation – already regulated in 1997 – into the civil code [strengthening housing-finance guarantees], the separation of project-based housing finance from overall company assets [in order to streamline asset recovery in bankruptcy processes] and the regulation of initial public offerings (IPOs) on capital markets for building and construction companies) (Royer, 2014). The latter led to a significant growth of IPOs of construction companies during 2006–2009 at the São Paulo stock exchange, and an increasing penetration of international investors in the Brazilian real estate sector, which, since the military regime, had been operating within a relatively protected national market. Developers and investors favoured a pro-active approach structured around early acquisition of land and the creation of land banks in order create financial founder's profits in non-metropolitan areas (Shimbo, 2012; Fix, 2011; Sanfelici, 2013). As it turned out, however, the international subprime crisis shook these excessively optimistic projections and generated a real threat of ending up with illiquid and high-risk investments in land and real estate. Nevertheless, the national government was quick to react to the international crisis and its effects on Brazilian spaces when it launched its anti-cyclical program My House, My Life (MHML) in 2009, which was both labour intensive and based on cheap credit to producers and consumers of housing.

The MHML program will be analyzed with more detail in Chapter 6, including a case study of its impact in the ABC region (a subset of seven cities located in the industrial heartland in the southeastern part of the metropolitan region of São Paulo). On that occasion we will discuss how the opportunity of a growing low-to medium-income market niche funded by the FGTS and HFS (targeted at families earning up to ten minimum salaries)[25] was quickly captured by the real estate industry. This led to escalating land and real estate prices and financial profits, which signalled a clear move away from the socio-economic segments originally targeted by the National Housing strategy itself. In the meanwhile, the private sector packaged this effective response to the program in terms of a discourse of having successfully unlocked a traditionally restrained market for low-income housing.[26]

3.5 Conclusion

The historical overview of this chapter allows us to draw some preliminary conclusions regarding the imbricated relations between the Brazilian state, cities and the production of space in times of re-emerging global finance.

First, despite the local availability of loanable funds, the country's trajectory of industrialization and urbanization has, by and large, occurred without consolidated and well-functioning domestic capital markets. Brazil has set these transitions in motion without building up significant markets for long-term credit, risk

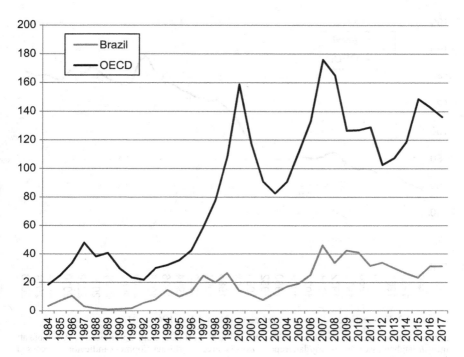

Figure 3.1 Stocks traded, total value (% of GDP, 1984–2017)

Source: Author's elaboration. Based on The World Bank. World Development Indicators. Available at: https://databank.worldbank.org/data/reports.aspx?source=world-development-indicators (accessed June 13, 2019).

The value of shares traded is the total number of shares traded, both domestic and foreign, multiplied by their respective matching prices. Figures are single counted (only one side of the transaction is considered). Companies admitted to listing and admitted to trading are included in the data. Data are end-of-year values.

capital and shares in order to finance its urbanization and industrialization. Some figures illustrate the historical analysis of this chapter. For instance, Figure 3.1, elaborated on the basis of the World Development Indicators from the World Bank, shows that the total value of shares traded at the stock exchange (as a percentage of GDP) has been consistently below the average figures for the OECD region for the period 1984–2017. Similar indicators on the supply of domestic credit to the private sector (Figure 3.2) and the provision of domestic credit by the financial sector (Figure 3.3, all in terms of percentages of the GDP), created on the basis of the same data source, illustrate at best a limited role of Brazilian capital markets as compared with the group of OECD countries.

Second, the Brazilian development project has been financed through a combination of explicit state efforts, international capital and, particularly in family-owned domestic groups, re-invested profits. The entanglements between the state and national companies have been particularly clear in traditionally sheltered segments such as building, construction and large contractors which, at least until

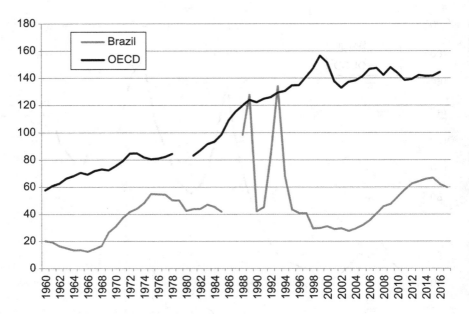

Figure 3.2 Domestic credit to private sector (% of GDP, 1960–2016)

Source: Author's elaboration. Based on The World Bank. World Development Indicators. Available at: https://databank.worldbank.org/data/reports.aspx?source=world-development-indicators (accessed June 13, 2019).

Domestic credit to private sector refers to financial resources provided to the private sector by financial corporations, such as through loans, purchases of nonequity securities, and trade credits and other accounts receivable, that establish a claim for repayment. For some countries these claims include credit to public enterprises. The financial corporations include monetary authorities and deposit money banks, as well as other financial corporations where data are available (including corporations that do not accept transferable deposits but do incur such liabilities as time and savings deposits). Examples of other financial corporations are finance and leasing companies, money lenders, insurance corporations, pension funds and foreign exchange companies.

recently, were characterized by a significant participation from family-owned businesses. Campos (2012), for example, analyzes in detail the linkages between the state and large constructors, which were being built up in a relatively protected domestic market during the military regime. At the same time, cities have accumulated chronic social-spatial deficits for lower-income groups in light of the historically selective entrance of the private sector – focussed on financially viable projects with short paybacks – and the lack of "really existing" Brazilian Spatial Keynesianism, which would have complemented the social wages with subsidized budgetary sources for housing and urban infrastructure. In the Brazilian setting, the public fund was never designed according to this modernist infrastructural ideal of connecting dynamic and stagnant regions within a national space economy, while reducing intra-urban disparities through predictable, massive budgetary support for affordable and well-located housing and urban services. If,

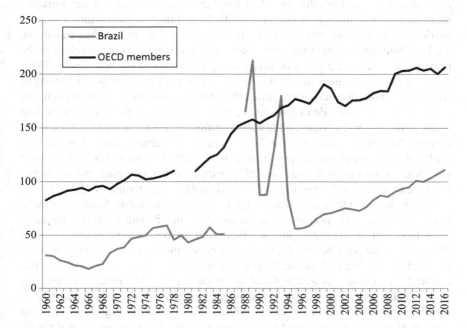

Figure 3.3 Domestic credit provided by financial sector (% of GDP, 1960–2016)

Source: Author's elaboration. Based on The World Bank. World Development Indicators. Available at: https://databank.worldbank.org/data/reports.aspx?source=world-development-indicators (accessed June 13, 2019).

Domestic credit provided by the financial sector includes all credit to various sectors on a gross basis, with the exception of credit to the central government, which is net. The financial sector includes monetary authorities and deposit money banks, as well as other financial corporations where data are available (including corporations that do not accept transferable deposits but do incur such liabilities as time and savings deposits). Examples of other financial corporations are finance and leasing companies, money lenders, insurance corporations, pension funds and foreign exchange companies.

paraphrasing Oliveira (1988), the public fund under Spatial Keynesianism guaranteed an indirect salary to workers through the *ex ante* subsidization of housing and urban infrastructure, this structural mechanism was never materialized in the Brazilian setting.[27] The FGTS that underpinned the housing finance arrangement of the Brazilian development state represented a compulsory savings scheme, charged from workers, which was based on the premise of cost recovery. At the same time, the selective involvement of the state in the constitution and preservation (in times of recurrent crisis) of subsidized credit for urbanization and industrialization led to the accumulation of a stock of government debt, which emerged as an attractive safe haven for domestic loanable funds and international capital in search of financial profits.

Third, the accumulation of the previously mentioned historical social deficits in cities and metropolitan areas has transformed them in privileged spaces to absorb and redirect the available domestic loanable funds into financial founder's profits, more particularly through speculation with land, urban infrastructure and public debt. In that sense, already in the initial stages of its rural-urban transition, the Brazilian development trajectory has witnessed what we have labelled in the previous chapter as a process of emerging urbanization of finance. Costa (2012: 428), for example, argues that "the bulk of the wealth of the "newly rich" emerged from non-productive activities, generally linked to mercantile valorization, for example, through the sale of real-estate, farms and shares". Moreover, according to the same author, considering the concentration of wealth in cities, it is no coincidence that "the majority of asset bubbles in Brazil occur in real estate markets (rather than the stock exchange)" (Costa, 2012: 429). In the empirical chapters of the book, we will analyze examples of how cities, urban planning and the public fund are involved in the making of financial profits in contemporary Brazil.

Fourth, the simultaneous availability of domestic loanable funds (that have been channelled into speculative urban founder's profits and securities linked to state debt) and the persistent inflow of external savings only represent a contradiction from the perspective of orthodox monetary theory. According to the latter, international savings fill in the "gap of domestic resources" in order to finance national investments, as well as fiscal and trade deficits. The Brazilian trajectory shows that, rather than (national and international) savings, it is the availability of speculative investment opportunities structured around government debt and urban founder's profits that have driven the circulation of money and credit. In that sense, the consistent accumulation of social deficits in cities and metropolitan areas and a stock of state debt and associated profitable securities have been instrumental in the absorption of national and international savings in search of financial profits in conditions of relatively low risk. The availability of these savings is both cause and consequence of the peculiar financial pattern that has underpinned the Brazilian production of urban space.

Fifth, the Brazilian public fund has never represented the *ex-ante*, predictable and stable complement to social wages in order to guarantee the reproduction of labour, as was prevailing under Spatial Keynesianism in many countries of the Global North. Indeed, rather than an incomplete version of these funding premises, Brazilian developmental finance has always been based on *ex-post* cost recovery through the contributions of a hybrid public fund sourced by compulsory contributions from labour, which have occasionally been complemented by ad-hoc support from the general budget, either in case of a crisis (for instance, after the bankruptcy of the National Housing Bank in 1985), or during very specific and exceptionally redistributive moments, such as occurred during the social-developmental regime under the Worker's Party administrations of 2003–2015. As mentioned, in Chapter 6 we will discuss this administration's flagship social housing program, MHML.

Sixth, this contradictory presence of the state and the public fund in the overall planning, finance and crisis management of the Brazilian developmental project of industrialization and urbanization has simultaneously, as a paradox, transformed fiscal austerity and reduction of governmental deficits and debts into one of the preferred narratives and objects of the neoliberal project aimed at transforming urban space and life itself into a set of tradable income-yielding assets. To be specific, in the Brazilian historical setting that has traditionally been marked by macroeconomic instability as well as high and fluctuating inflation levels and interest rates, state debt has been one of the preferred and traditional safe havens for the absorption of domestic loanable funds and international savings. Therefore, in the particular Brazilian context it becomes increasingly important to investigate the reasons why, as well as the emerging financial and institutional devices, narratives and governance arrangements that accompany the continuous reframing and restructuring of fiscal austerity and the nexus between cities and the public fund in emerging capital markets. Brazilian metropolitan areas are prime case material, considering that their land and infrastructure networks have always been privileged platforms for the absorption of loanable funds, not only from smaller cities and rural areas (which have accumulated successive trade deficits with metropolitan areas), but as well as through the inflow of international savings (Dymski, 1998). The restructuring of the city-capital market nexus under austerity in the contemporary Brazilian setting will be explored in more detail in Chapter 7.

Notes

1 Although it was prevailing until 1930, its gradual transformation had already been set in motion since the First World War, driven by the emergency substitution of interrupted imports as well as by successive balance of payment problems and exchange rate crises.
2 As a colony, these remittances were initially directed to the Portuguese crown. In a subsequent stage, profits were increasingly remitted to London as the emerging global financial centre.
3 This section is based on the work of Fundação Perseu Abramo (2019) and Costa (2012).
4 Formal independence was proclaimed in 1822, but the Portuguese court had been relocated to Brazil and the proclamation of the republic only occurred in 1889.
5 Somewhat different from São Paulo, the banking system of the capital city of Rio de Janeiro became specialized in the emission and management of state debt and government securities.
6 The exchange rate stabilization mechanism of 1906 was approved by law (the so-called cash conversion mechanism or *caixa de conversão*) and would also prove essential to realize the government support to the coffee sector. Stabilization of the exchange rate was key in providing minimum price guarantees to domestic coffee producers. To be specific, considering that exporters receive international currency, a hypothetical appreciation of the Brazilian currency would threaten domestic debt repayment by coffee producers.
7 For a detailed conceptual and numerical discussion see also chapter 2. In the specific case under consideration, the dividends are capitalized with the relatively low (subsidized) interest rates that could be obtained in Brazil, generating a higher net present value and subsequent founder's profit by floating Brazilian companies in European stock exchanges.
8 This measure would also prove instrumental in avoiding that foreign banks would continue to use the domestic market as a privileged space of generating financial profits through gearing, thereby minimizing the need to mobilize foreign investment funds.
9 In addition to the broader objective of coordinating more closely monetary, credit and exchange rate policies, the need for a stricter regulation of German, Japanese and Italian banks also emerged in light of these countries' role as enemies in the Second World War.

10 It continues to perform a similar role under a slightly different name, that is, the National Bank for Economic and Social Development (*Banco Nacional para o Desenvolvimento Econômico e Social*).

11 A first best solution would be the elimination of inflation itself, which was only realized at the time of the Plano Real in 1994.

12 Currency boards imply reduced domestic discretionary power over monetary policy considering the fact that national currencies are pegged to key money such as the dollar. Brazil's Plano Real never went that far since it remained the scope for policy discretion over the exchange rate in light of international shocks, such as the one that occurred in 1998. Argentina, however, inserted the clause that pegged the exchange rate of the peso to the dollar into its constitution. This generated a dramatic scenario whereby domestic living conditions were gradually adjusted downwards according to the successive external shocks that affected the country during 1998–2001. It ultimately led to the December 2001 street riots and the resignation of president de la Rúa after the restrictions on people's ability to withdraw cash from banks (*the Corralito* in Spanish).

13 This section is a based on Klink and Denaldi (2014: 221–223).

14 Autonomous savings and credit schemes without state regulation and intervention had existed since 1910.

15 Two lines of operations were targeted at rental and/or owner-occupied housing for members while a third introduced mortgages to external clients.

16 The distant housing estates that had initially been constructed without much infrastructure implied a subsequent rollout of infrastructure finance (Bolaffi, 1977).

17 Moreover, loan repayments were indexed on a yearly basis while monetary correction of outstanding balances occurred quarterly, thus generating additional implicit subsidies and financial deficits within the system.

18 FCVS, the literal translation being the "Fund aimed at compensating the variation in growth of salaries" (*Fundo de Compensação de Variação Salarial*). Estimates by the Ministry of Finance in 2009 amounted to a present value of debt of around R$ 170 billion.

19 Depending on the original source of funds the coupon rate would be 3.12% or 6% plus inflation correction.

20 PROER stands for *Programa de Estímulo à Reestruturação e ao Fortalecimento do Sistema Financeiro Nacional.*

21 For a paradigmatic example of the influential international circulation of ideas on market-friendly housing finance see the international seminar, "Housing Finance in Emerging Markets: Policy and Regulatory Challenges", which was organized by the World Bank Group during March 10-13, 2003 in its head-office in Washington. It was explicitly targeted at officials working in the "old" developmental housing bank institutions in countries of the Global South, such as Brazil, India, Thailand and Indonesia.

22 The so-called Fundo de Garantia de Tempo de Serviço (FGTS, meaning Guarantee Fund Based on Service Time Contributions) would become the main source of low-income housing finance.

23 Finally, 20% of all savings were to be deposited at the central bank, while 15% was to be held as a liquidity reserve.

24 Most of the system's initial operating experience was achieved in the commercial sector where a more stable and predictable cash flow profile provided a relatively superior basis for project-finance.

25 In January 2019, a minimum salary is R$ 998,00, that is, 247,64 USD (exchange rate of 4.03 (14/8/2019).

26 The "really existing" low-income market is composed of families earning up to three minimum salaries.

27 This is what Oliveira (1988) labelled a state-mediated generation of anti-value. More specifically, the state used the public fund to contribute to the formation of a social wage through the subsidization of housing and urban infrastructure, which were

removed from the mechanism of capitalist reproduction and re-distributed to workers. As such, the welfare state was directly involved in the reduction of risk for households and labourers. See Chapter 2.

References

Aglietta M (2018) *Money: 5,000 years of debt and power.* London and New York: Verso.

Araújo TB de (1999) Por uma política nacional de desenvolvimento regiona. *Revista economica do Nordeste* 30(2): 1–30.

Arretche MT (1996) Desarticulação do BNH e autonomização da política habitacional. In: Affonso R de BA and Silva PLB (eds.) *Descentralização e políticas sociais.* São Paulo: FUNDAP, pp. 107–138.

Bielschowsky R (1988) *Pensamento econômico brasileiro. O ciclo ideológico do desenvolvimentismo.* Rio de Janeiro: Contraponto.

Bolaffi G (1977) *A casa das ilusões perdidas: aspectos sócio-econômicos do Plano Nacional de Habitação.* São Paulo: Brasiliense/CEBRAP.

Bonduki N (1998) *Origens da habitação social no Brasil.* São Paulo: Estação Liberdade.

Brasil (2010) *Plano Nacional de Habitação.* Brasília: Ministério das Cidades.

Brenner N and Elden S (eds.) (2009) *State, space, world: Selected essays.* Henri Lefebvre. Translated by Gerald Moore; Neil Brenner and Stuart Elden. Minneapolis: The University of Minnesota Press.

Cagnin RF (2012) A Evolução do financiamento habitacional no Brasil entre 2005 e 2011 e o desempenho dos novos instrumentos financeiros. *Boletim de Economia* 11(1): 15–32.

Campos PHP (2012) *A ditadura dos empreiteiros. As empresas nacionais de construção pesada, suas formas associativas e o Estado ditatorial brasileiro, 1964–1985.* Tese de Doutorado. Universidade Federal Fluminense, Instituto de Ciências Humanas e Filosofia, Departamento de História, Niterói.

Cano W (1998) *Desequilíbrios regionais e concentração industrial no Brasil, 1930–1995.* Campinas: Instituto de Economia da UNICAMP.

Costa FN da (2012) *Brasil dos bancos.* São Paulo: EDUSP.

Dymski GA (1998) "Economia de bolha" e crise financeira no Leste Asiático e na Califórnia: uma perspectiva espacializada de Minsky. *Economia e Sociedade* 11: 73–136.

Eloy CM de M; Costa F de C and Rossetto R (2013) Subsídios na política habitacional brasileira: do BNH ao PMCMV. In: ANPUR (ed.) *Encontros Nacionais da ANPUR.* Recife: ANPUR, pp. 1–20.

Fernandes AC and Novy A (2010) Reflections on the unique response of Brazil to the financial crisis and its urban impact. *International Journal of Urban and Regional Research* 34: 952–966.

FGV Fundação Getúlio Vargas (2007) *O Crédito Imobiliário no Brasil.* São Paulo: FGV Projetos.

Fix M de AB (2011) *Financeirizacão e transformações recentes no circuito imobiliário no Brasil.* Campinas: Universidade de Campinas.

Fundação Perseu Abramo (2019) *Os donos do dinheiro. O rentismo no Brasil.* São Paulo: Fundação Perseu Abramo.

Furtado C (2005) *A formação econômica do Brasil.* São Paulo: Companhia Editora Nacional.

Granja PRS (2008) *Modelo para avaliação do custo burocrático do FCVS: Um estudo de caso.* Dissertação de mestrado. Escola brasileira de administração pública e de empresas, Fundação Getúlio Vargas, Rio de Janeiro.

Hilferding R (1910) *Finance capital*. London: Routledge & Kegan Paul (1981 Edition).

Klink J (2013) Development regimes, scales and state spatial restructuring: Change and continuity in the production of urban space in metropolitan Rio de Janeiro, Brazil. *International Journal of Urban and Regional Research* 37: 1168–1187.

Klink J and Denaldi R (2014) On financialization and state spatial fixes in Brazil: A geographical and historical interpretation of the housing program My House My Life. *Habitat International* 44: 220–226.

Klink J and Keivani R (2013) Development as we know it? Change and continuity in the production of urban and regional space in Brazil. *International Journal of Urban Sustainable Development* 5(1): 1–6.

Maricato E (2011) *O impasse da Política Urbana*. Rio de Janeiro: Editora Vozes.

Martins J de S (2010) *O cativeiro da terra*. São Paulo: Contexto.

Mello MABC (1988) Classe, burocracia e intermediação de interesses na formação da política de habitação. *Revista Espaço e Debates* 24: 75–85.

Ministry of finance/Secretaria do Tesouro Nacional (2015) *Novação de dívidas do FCVS e Assunção de Dívidas pela União*. Brasilia: Ministry of Finance.

Miranda JC and Tavares M de C (1999) Brasil: estratégias de conglomeração. In: Fior JL (ed.) *Estados e Moedas no desenvolvimento das nações*. Petrópolis, RJ: Vozes, pp. 327–350.

Myrdal G (1972) *Teoria Econômica e regiões subdesenvolvidas*. Rio de Janeiro: Saga.

Oliveira F de (1981) *Elegia para uma re(li)gião: SUDENE, Nordeste, Planejamento e conflito de classes*. 3rd Edition. Rio de Janeiro: Paz e Terra.

Oliveira F de (1988) O surgimento do Antivalor. Capital, força de trabalho e fundo público. *Novos Estudos* 22: 8–28.

Oliveira F de (1990) A metamorfose da arribaçã: fundo público e regulação autoritária na expansão econômica do Nordeste. *Novos Estudos CEBRAP* 27: 67–92.

Paulani LM (2008) A Crise do regime de acumulação com dominância da valorização financeira e a situação do Brasil. *Estudos Avançados* 23(66): 25–39.

Prado Jr C (1994) *História Econômica do Brasil*. São Paulo: Brasiliense.

Rangel I (1981) A história da dualidade brasileira. *Revista da Economia Política* 1(4): 5–34.

Royer L (2014) *Financeirização da Política Habitacional: Limites e Perspectivas*. São Paulo: FAU-USP.

Sanfelici D (2013) Financeirização e a produção do espaço urbano no Brasil: uma contribuição ao debate. *Eure* 39: 27–46.

Shimbo LZ (2012) *Habitação social de Mercado*. Belo Horizonte: Editora Arte.

Soja EW (1989) *Postmodern geographies: The reassertion of space in critical social theory*. London: Verso.

Tavares M de C (1999) Império, Território e dinheiro. In: Fior JL (ed.) *Estados e Moedas no desenvolvimento das nações*. Petrópolis, RJ: Vozes, pp. 449–489.

Topik S (1979) Capital estrangeiro e Estado no Sistema Bancário Brasileiro: 1889–1930. *Revista Brasileira de Mercado de Capitais* 5(15): 395–421.

Valença MM and Bonates MF (2010) The trajectory of social housing policy in Brazil: From the National Housing Bank to the Ministry of Cities. *Habitat International* 34(2): 165–173.

4 Planning, projects and profitability

The role of financial and institutional devices in the urbanization of finance[1]

4.1 Introduction

This chapter is the first in a row that provides empirical illustrations on the entanglements between cities, urban planning and management in the making of urban financialization (Halbert and Attuyer, 2016). More specifically, we will flesh out the effective design and operationalization of large urban redevelopment projects through the securitization and sale in public auctions of additional building rights (CEPAC)[2] in urban public-private partnerships (Operações Urbanas Consorciadas – Urban Partnership Operations (UPOs)). Brazilian federal legislation (City Statute) enabled cities to implement, within specific perimeters established of the master plan, urban redevelopment projects that are financed by charging developers for the provision of additional building rights. While these charges can be implemented through formula-based mechanisms (*outorga onerosa*), cities such as São Paulo have experimented intensively with CEPACs in order to capture the acclaimed advantages of the capital market that, according to some of the rather "self-referential" interviews we conducted with financial intermediaries, represents part of "the emerging Brazil".[3]

UPOs financed by CEPACs have similarities with tax increment finance districts that have been disseminated widely in the North-American context (Weber, 2010). They also allow, at least theoretically, the generation of upfront financial resources from projected urban redevelopment, thereby capitalizing expected additional revenues within predefined perimeters of project areas. Unlike TIF, however, CEPACs are *not* part of municipal debt and can be considered tradable "non-interest-yielding assets", which can be acquired by financial investors, commercialized in secondary markets or just used by builders and developers in the perimeter of projects. Moreover, as will be explained in this chapter, CEPACs can serve as outright credit money in order to pay contractors and developers or to buy into real estate investment funds, among other possibilities, under the condition this contributes to achieving the objectives of particular urban redevelopment projects.

Our analysis will focus on the discrepancies between idealized and "really existing" UPOs with CEPACs by investigating the circulation of norms, calculative practices and pricing conventions that underpin the financial and physical design of large urban redevelopment projects. The official narrative behind the instrument is grounded in orthodox urban economics and finance (DiPasquale and Wheaton, 1996). The key element is the pricing of CEPACs, which is supposed to perfectly connect competitive property (land owners, construction firms, contractors, developers) and financial asset markets for the built environment (potential investors), as well as provide local governments with the required amount of resources in order to finance infrastructure as well as public goods such as low-income housing (Blanco et al., 2016). Really existing UPOs with CEPACs, however, are designed in a context of unconsolidated capital markets, with local stakeholders not merely performing the role of "price-takers", but actively shaping the financial and physical design and implementation of urban redevelopment projects. Moreover, although the absence of secondary asset markets still indicates limits to the penetration of financial metrics into urban planning, the instrument's open-ended and dynamic trajectory, marked by a continuous evolution of financial and institutional devices, nevertheless suggests ongoing and variegated pressures on the transformation of urban space into an asset.

This chapter is a result from the work that was undertaken by Klink and Stroher (2017), which was based on a review of the literature, research of files and official documents from UPOs and 17 semi-structured interviews. It extends on these efforts by including a closer look at the financial-institutional devices that are behind a new generation of urban redevelopment projects in São Paulo that are emerging in the post-2015 scenario of economic crisis.

After this introduction, the chapter is organized in four complementary sections. The first provides a general overview of the characteristics as well as the narratives behind the dissemination of UPOs financed by CEPACs. The second section analyzes "really existing" UPOs, while the third presents a preliminary reading of some of the institutional-financial devices that have accompanied the emergence of a new generation of urban redevelopment projects in São Paulo through the so-called Urban Intervention Projects (i.e. o *Projeto de Intervenção Urbana* (PIU)). The chapter is concluded with suggestions for further research.

4.2 A primer on UPOs, developer contributions and CEPACs

Brazil has experimented with developer contributions since the 1970s (Maleronka, 2010). São Paulo, for example, created its *outorga onerosa* and *solo criado* (additional land creation). The latter stipulated that developers would have to pay contributions whenever the built-up area exceeded the size of the land. Local governments could use revenues to provide parks, infrastructure and low-income housing.

In the 1980s and 1990s Rio and São Paulo also worked with *operações interligadas* (linkage operations), whereby owners with slums on their properties could develop beyond predefined floor area ratios, provided they offered low-income housing in other (linked) areas. Considering that linkage operations prioritized areas attractive to the market, they proved disappointing in light of their (theoretically) redistributive objectives (Fix, 2009).

It was not until the approval of the federal City Statute (CS) in 2001 that this praxis with developer contributions became part of a broader institutional framework for urban reform. Decentralization and re-democratization of the 1980s set the stage for the re-emergence of social movements, elected mayors and progressive planners that all denounced the contradictory trajectory of cities and demanded better living conditions. This pressure led to an urban chapter in the 1988 constitution that was framed around collaborative-participatory master planning and the social function of private property, operationalized through the CS by a series of redistributive land market instruments such as low-income zones; compulsory subdivision, utilization and building on vacant land; and progressive property taxes over time. (Ondetti, 2016)

Simultaneously, real estate lobbies succeeded in negotiating UPO financed by developer contributions into the CS. UPOs require regularization within the master plan that is approved by the city council. This includes a detailed description of perimeters, the list of infrastructure investments and broader objectives of operations. CEPACs are tradable non-interest-yielding assets and, in certain circumstances, represent credit money. For example, local governments are allowed to use CEPACs as compensation in expropriations of landowners or as payments to contractors (if these accept). Likewise, CEPACs can be used to acquire participation in real estate investment funds or mixed-capital specific-purpose companies that can be set up by local government in order to reach the objectives of UPOs.

In theory, the pricing of CEPACs in UPOs articulates users in land and property markets, investors in asset markets and local government. Constructors-developers buy CEPACs in public auctions in order to obtain additional building rights, specified for each district within UPOs on the basis of the difference between basic and maximum floor area ratios. Potential investors acquire CEPACs looking for capital gains in successful operations. The success of UPOs with CEPACs is intrinsically linked to real estate valorization. More particularly, the upfront revenues associated with the initial sale of part of the stock of CEPACs enables local government to implement the initial part of projected infrastructure investments, which valorizes the area. Subsequently, builders and developers perceive the momentum of a particular UPO and are willing (and able) to offer higher prices for CEPACs in future auctions. While providing capital gains to investors and positive expectations regarding future appreciations in the asset market, higher revenues associated with the sale of CEPACs may be recycled back into the area by local government through investments in social housing, infrastructure and other local public goods.

A selective literature has provided arguments for the advantages of CEPACs as a market-oriented source of finance for UPOs over formula-based development charges. (Sandroni, 2010; Maleronka, 2010; Smolka, 2013).

First, auctions of CEPACs guarantee upfront resources for local governments. This is preferable to project-by-project cash inflows received through development charges, whereby revenues depend on an efficient approval and implementation of proposals.

Second, auctioning CEPACs stimulates competitive bidding among developers and potential investors, thereby leveraging local government revenues that can be recycled back into the area.

Third, different from formula-based development charges, the sale of CEPACs provides a virtuous cycle between land and property and asset markets, whereby positive expectations regarding the trajectory of UPOs are capitalized into the value of the area and the prices of certificates, triggering both investor interest and additional city revenues.

Fourth, resource mobilization through CEPACs represents land-value capture with advantages in terms of distributional justice. More specifically, considering a fixed supply of land and the possibility of user-developers to substitute neighbouring plots, if well designed and implemented, the burden of financing local public goods through CEPACs is neutral and will ultimately fall on property owners. Higher floor-area ratios in UPOs (increasing land prices) and expenses with CEPACs (lowering them) are expected to cancel large effects on land prices.

Finally, CEPACs require procedures aimed at providing transparent urban governance and investor protection, and their emission in public auctions is supervised by the Brazilian equivalent of the Securities and Exchange Commission (BOVESPA). After approval of the UPO by the municipal council, local governments must publish a prospectus providing investor information on infrastructure investments, financial-economic viability, risks and environmental impacts.

However, some of these advantages have been contested. For example, "really existing" UPOs are believed not to overcome a built-in tension between the requirement for rapid valorization of project areas and CEPACs and the need to provide social housing, which tends to depreciate market values of real estate. (Fix, 2009; Coma, 2011; Cardoso, 2013; Siqueira, 2014) Likewise, the use of UPO with CEPACs is considered to have led to excessive emphasis on infrastructure finance and revenue generation, triggering a disproportional increase in size and scale of operations as compared to international UDPs (Maleronka, 2010).[4] Moreover, claims of transparency associated with capital markets seem exaggerated. The numerous reports that accompany UPOs with CEPACs are targeted at builders/developers and investors, positioning them around the likely risks and returns of the instrument. Considering the highly technical language used, these reports are unlikely to be of much use to increase the understanding of the average citizen and inhabitants regarding the design, implementation and impacts of the instrument. (Salomão, 2016) In the meantime, the municipality of São Paulo has published implementation reports about development contributions that are similar to the information provided by the capital market on CEPACs. Finally, as will be analyzed, the pricing of CEPACs doesn't reflect competitive bidding and, in practice, is influenced by a select group of planners and financial consultants responsible for viability studies and developers that drive the contradictory design and implementation of UPOs (McAllister, 2017).

4.3 UPOs with CEPACs at work

Somewhat different from the orthodox narrative, the market for CEPACs has not been characterized by price-takers, whereby prices would provide the correct incentives to users and investors, while providing social infrastructure and local public goods. In practice, the pricing of certificates has been driven by the circulation of norms, expectations and metrics within a community of consultants, urban planners, builders, developers and contractors structured around the design and implementation of projects that would guarantee financial viability to builders-developers, while providing local government a sufficient amount of revenues in order to finance urban infrastructure that could be built by contractors. In this complex setting, it has proven challenging to match these often-contradictory expectations and to create liquid investor markets for CEPACs.

4.3.1 Design

In Table 4.1, which was elaborated on the basis of an analysis of prospectuses, viability studies and reports from local governments and consultancies, we summarize parameters of UPOs using CEPACs in metropolitan Rio, São Paulo and Curitiba (PMC (Prefeitura Municipal de Curitiba) (2015); PMSP (Prefeitura Municipal de São Paulo) (2014b); PMSP/SP Urbanismo (2015b)).

Espraiada combines investments in roads, public transportation, social housing and public space. *Faria Lima* is designed around investments in roads, improved connectivity between train and metro and slum upgrading. *Porto Maravilha* in Rio de Janeiro is structured around the recycling of downtown port areas through land-readjustment, investments in housing, roads and urban public space. *Linha Verde* is aimed at up-scaling Curitiba's well-known rapid bus system integrating mobility, land-use and urban space. The objective of São Bernardo's UPO is to improve urban transportation and infrastructure.

A university and a private financial firm attend the small consultancy market for viability studies. The latter concentrates the bulk of the work and has considerable track experience with valuation since the 1990s when São Paulo experimented with *solo criado*, development charges and CEPACs before the enactment of the CS. It also works with cities on UPOs that are in the pipeline, such as Belo Horizonte, Niteroi and São José dos Campos, among others.

Contracted by local government, the consultant's work represents a balancing act in terms of providing a price of CEPACs that provides a perspective of financial viability for builders and developers, attractiveness for potential investors and sufficient revenues for cities.

All viability studies we investigated are grounded in academic financial economics and discounted cash flow models whereby projected net revenues are capitalized to present values so as to provide an indicator of the maximum amount of money that can be spent on both land and CEPACs in order to maintain viable rates of return compared with the prevailing cost of capital. Moreover, the owner of the private consultancy we interviewed uses his long-established networks

Table 4.1 Urban Partnership Operations (UPOs): Main physical and financial design parameters (real prices – base year first prospectus) (*)

1	2	3	4	5	6	7	8	9	10	11	12	13
Name of Intervention	Base year first prospectus	Extra Building Stock (m2)	Quantity of certificates	Consultancy firm Viability study		Discount Rate	"Implicit" IRR/NPV (*****)	Viable Minimum unit price CEPACs (R$)	Minimum Unit price of CEPACs in prospectus	Expected Min. Sales Revenues (R$)	Infrastructure costs of operations (R$)	Objectives UPO
SBC – Centro	2015	2,847,000	4,967,000	Amaral D'Avila		16%	na	R$ 400	na	R$ 1,986,800,000	nd	Strengthening of centrality in areas of interest to the real estate market
Rio – Porto Maravilha	2012	4,089,502	6,436,722	Amaral D'Avila		15%	between 21%–34%	R$ 580	R$ 545	R$ 3,508,013,490	R$ 8,201,332, 019	Renovation of old inner-city port area
Curitiba – Linha Verde	2012	4,475,000	4,830,000	FIPE		18%	between 12%–82%	between R$ 598–R$ 928	R$ 200	R$ 966,000,000	R$ 1,200,000,000	Renovation of areas next to Rapid Bus-Highway system
São Paulo (SP)/Faria Lima	2004 (**)	2,250,000	1,000,000	Amaral D'Avila		18%	na	between R$ 3,390–R$ 10,207	R$ 1,538	R$ 1,538,000,000	R$ 2,230,000,000	Creation of centrality in areas of interest to the real estate market

(Continued)

Table 4.1 (Continued)

1	2	3	4	5	7	8	9	10	11	12	13
SP/Água Branca – Res.	2014 (***)	1,350,000	1,605,000	FIPE	17%	NPV= 17,5% of Gross Sales Value	Average R$ 1,665	R$ 1,548	R$ 2,484,540,000	R$ 4,954,358,938	New Real Estate frontier, connection to rail and highway network
SP/Água Branca – N. Res		500,000	585,000				Average R$ 1,666	R$ 1,769	R$ 1,034,865,000		
SP/Água Branca – Total		1,850,000	2,190,000				Average R$ 1,667		R$ 3,519,405,000		
SP/Água Espraiada	2004 (****)	3,750,000	3,750,000	Amaral D'Avila (2004) FIPE (2008)	16% 22%	na na	between R$ 402–R$ 726 between R$ 477–R$ 807	R$ 300 R$ 300	R$ 1,125,000,000	R$ 1,255,000,000	Creation of centrality in areas of interest to the real estate market

Source: Author's elaboration. Based on viability studies and prospectuses: PMC (2012), PMRJ (2012), PMSBC (2015), PMSP (2008, 2012, 2014a).

(*) Exchange rate: 1USD= R$ 3.16. 25/9/2017
(**) Updated in 2012
(***) Started as an Urban Operation in 1995
(****) Updated in 2008
(*****) NPV = Net Present Value. IRR= Internal Rate of Return

and approaches "the market" regarding the viability of certain price ranges for CEPACs:

> I am not an academic. Many builders and developers have their projections, required rates of return and price ranges. Some are not willing to open this up so you have to provoke them by throwing them some of these figures.
>
> (Interview no. 2, Financial Consultant, November 2016)

In Table 4.1 we traced this recursive relationship with the "market" in performing viability studies. To facilitate comparisons between the projections of viability studies and the prices of CEPACs effectively published in the prospectus, all financial data in Table 4.1 are in real values of the base year the first prospectus was launched.

The initial pricing of CEPACs establishes comfortable projected rates of return under the worst scenario conditions (columns 7 to 10). In column 7 we show the interest rates prevailing in the base year of each UPO. This variable provides key information to the market: if the internal rate of return (IRR) (column 8) is higher than interest rates or, alternatively, if capitalization of net project revenues (revenues minus costs) at that same rate provides a positive net present value (NPV), there is a clear perspective of financial viability. Both the UPOs in Rio and Curitiba were designed to provide an IRR that exceeded the prevailing cost of capital under scenarios of cost overruns.[5] The simulations behind *Branca* were even more explicit in pre-fixing a NPV of capitalized net revenues equal to 17.5% of the projected gross sales value (column 8).

Column 9 shows that this financial engineering around predetermined economic returns has implications for the pricing of building certificates. The minimum price is an initial parameter used in public auctions: higher prices tend to increase public revenues but reduce developer's profitability. Therefore, there is a relationship between the IRR and NPVs (columns 7/8) and the minimum prices of CEPACs that are financially viable for developer/builders (column 9). Moreover, the values shown in column 9 represent "worst-case scenarios", considering they are based on calculations performed on the basis of cost-revenue projections in the least profitable districts within UPOs, thereby reducing risks for builders and developers.

Column 10 illustrates the minimum price of CEPACs that was effectively used in the prospectus to start the emission process. With the exception of *Branca* (approximately at break-even level), where pricing was criticized by the market, minimum prices in the prospective were always set below the ranges simulated in the viability studies in order to provide favourable rates of return for developers and good perspectives for investor valorization. For example, while projections for *Linha Verde* pointed out that developers could afford to pay minimum prices between R$ 598,00–R$ 928,00 in the least attractive district of the UPO, the prospective consolidated a minimum price of R$ 200,00.

The financial consultancy stressed the relevance of initial conversations with local governments regarding the "optimal design" of UPOs in terms of pricing of CEPACs, expected public revenues, infrastructure investments and the size of operations. Increasing the perimeter of UPOs would introduce more complexity but also provided builders and developers with leverage over landowners by not depending on specific plots in smaller areas. Moreover, the increased stock of CEPACs would allow governments to design and implement a structured emission policy over time:

> On the basis of viability studies we initially provide a projection of the expected financial revenues for local government. If it wants more, this requires increasing the perimeter and the stock of CEPACs as well as chang-ing the list of investments.
>
> (Interview no. 2, Financial Consultant, November 2016)

At first sight, this sounds simple. However, a series of complexities emerge in the design of UPOs. For one, as can be seen in columns 11 and 12 of Table 4.1, in most UPOs the present value of projected sales revenue of CEPACs at minimum price are lower than the costs of public infrastructure. Rio de Janeiro's *Porto Mara-vilha* is emblematic of this "built-in" design with a projected R$ 3,5 billion of sales revenues (at a minimum price of R$ 545,00) against a total cost of infrastructure estimated at R$ 8 billion. The initial pricing of CEPACs locks local governments, de facto and de jure, into an arrangement of producing increasing user and inves-tor value for money in UPOs. One of the brokers summarized this crucial role of expectations, pricing and the "momentum" created by public works in the UPOs:

> Investor confidence depends on public works and the subsequent valorisation of the area. Subsequently, the market will start to "price" the area much better and a virtuous cycle will be set in motion.
>
> (Interview no. 3, Banco do Brasil, December 2016)

There is nothing inherent about this "virtuous cycle", however. As a planner observed:

> The minimum price of certificates, analysed isolated, does not say much; it doesn't necessarily correspond with the infrastructure investments of UPOs. As such, the Rio experience represented a big mistake. The market doesn't have the capacity to finance all that.
>
> (Interview no. 4, São Paulo City Planner, November 2016)

Moreover, some interviews revealed that this "optimization" of design had led to excessively large UPOs, which challenged an integrated management. As the same planner put it:

> The size of the average UPO in São Paulo and elsewhere is huge. This has happened because planners have aimed to create a redistributive "Robin Hood" type of mechanism for land market intervention.
>
> (Interview no. 4, São Paulo City Planner, November 2016)

In that sense, the design of the perimeter of the UPO in São Bernardo was emblematic; it distinguished between a "core perimeter", representing the prime real estate market where CEPACs could be bought and valorization was to take place, and an "extended perimeter" in the outskirts, containing one of the biggest slums called *Montanhão,* where 20% of revenues were earmarked to social housing and public transportation. According to the financial consultant, this "spatial fix" by extending the perimeter of UPOs in order to resolve the pressure for social housing was a logical move:

> This insistence by planners on mixed-income housing in prime locations will simply not work because the market will not buy into it.
> (Interview no. 2, Financial Consultant, November 2016)

Finally, some builders/developers complained about the expensive infrastructure standards that marked UPOs in light of contractors' influence over the design of UPOs:

> Because of the lobby of contractors, there is a bias in favour of large and expensive infrastructure such as roads and monorail as compared to public transportation and light rail.
> (Interview no. 5, Developer 1, November 2016)

4.3.2 Implementation

Public sale of CEPACs in auctions is supposed to trigger competitive bidding, thereby increasing prices and local government revenues, while stimulating the development of liquid and low-cost secondary markets. (Smolka, 2013)

Figure 4.1a illustrates how this bidding has worked out in the UPO AE (all prices in terms of 2013). With two exceptions, there has been no agio and the minimum prices of the prospectus coincide with effectively realized prices. While most interviews considered the agio of 2008 as "accidental", the second occurred considering the depletion of the stock of CEPACs at the end of the UPO.[6] Figure 4.1a also illustrates that realized prices in auctions were consistently below the values that had been calculated in viability studies around minimum guaranteed rates of return for developers, as discussed earlier.[7]

Figure 4.1b also illustrates that it has proven challenging for government to "outsmart" the market through an emission policy aimed at creating scarcity and increasing prices of CEPACs. The broker we interviewed confirmed that, after initial government efforts in that direction, developers-buyers organized in order to "overshoot" their revealed collective demand and prevented scarcity of CEPACs. In several occasions the supply of CEPACs exceeded market demand.

Making a market for CEPACs has proven complex. Preparing emissions requires contracting underwriters and brokers in order to mobilize buyers and investors. Considering the more flexible procedures, São Paulo ended up contracting a public institution, *Banco do Brasil,* to coordinate this work. Without

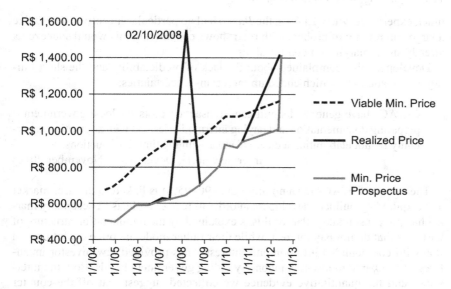

Figure 4.1a Public Auctions in UPO AE (2004–2012): per unit viable prices, min. prices in prospectus and realized prices of CEPACs (Real Prices: Dec. 2013 = 100)

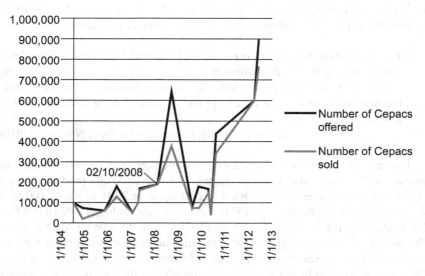

Figure 4.1b Public Auctions in UPO AE (2004–2012): number of CEPACs offered and sold

much experience with CEPACs, the *Banco* had no particular incentive to mobilize a large community of brokers with road shows considering this would squeeze its already small margins on commissions.

Developers also complained about the lack of predictability and the short timing before auctions, which ended up increasing uncertainties:

> CEPACs have generated significant transaction costs for local government – preparing documentation and paying intermediaries – as well as creating uncertainty by not elaborating a clear calendar with the timing of auctions.
>
> (Interview no. 5, Developer 1, November 2016)

The bulk of CEPACs (around more than 90–95%) is linked to the user market and acquired by builders/developers in order to realize projects. A secondary market has not consolidated. Part of this is explained by the remuneration structure of CEPACs that do not pay interest, while their rather moderate price evolution – in line with compounded inflation and interest – has provided few investor incentives as compared to low-risk income-yielding government bonds. Both the interviews and the quantitative evidence we collected suggest that off-the-counter transactions indicated a weakly developed secondary market, dominated by users who needed small quantities of CEPACs in between public auctions in order to complement projects, frequently paying higher prices than in public auctions.[8]

The lack of secondary markets and the smooth evolution of prices of CEPACs introduce additional complexity regarding the debates on the instrument's attributed effects on price escalation in land and real estate markets. As one of the consultants active in the dissemination of the instrument provoked:

> This critique on the correlation between CEPACs and speculation is confused. The secondary market that would allow these strategies is too small (3–5%), while during the rare occasion where speculation did occur – February 2008 – the particular actor made a bad bet and generated extra revenue for local government.
>
> (Interview no. 6, Consultant CEPAC, November 2016)

The lack of secondary markets doesn't exclude the instrument's hypothetical impact on the escalation of real estate prices, however. While product prices in *Agua Espraiada* increased from R$ 4,532.00/sq. m in 2004 to R$ 11,100.65 in 2013, the figures for the same period in the city of São Paulo indicated a rise from R$ 4,622.28 to R$ 7,857.51 (Centro de Estudos da Metrópole/CEM). Moreover, the combined effect of higher floor area ratios and public infrastructure investments in UPOs (increasing land prices) and expenses with CEPACs (depressing land prices) seem to have cancelled out, considering the fact that land prices in and around *Agua Espraiada* – up to a distance of 1 km – have stabilized around a level of R$ 2,000. (PMSP, 2015a) Independent of land prices, the bottom line is that housing prices in and around a border of 500 metres of the UPO are at a similar high level, and rising steeply. UPOs with CEPACs haven't democratized the right to well-located and affordable housing in cities.

The UPO *Porto Maravilha* in Rio de Janeiro seems to have launched a new phase in the design and implementation of UPOs with CEPACs. (Pereira, 2016) First, through a political alliance between the city, state and federal government, combined with regulatory flexibilization,[9] the federal housing bank used resources from the earlier mentioned Social Housing Finance system – funded by surcharges on formal sector wages – to buy the CEPACs that were offered in one non-divisible tranche in the first auction. This signalled one more step in a longer trajectory of regulatory restructuring around the reduction of the borders between finance for social housing and real estate (Royer, 2014). The bank then used its CEPACs to buy into a real estate investment fund with a direct stake in *Porto Maravilha*. This meant that CEPACs, non-interest yielding assets, were now leveraging the UPO by an increased circulation of tradable "income yielding" assets within its perimeter. Finally, Rio marked a change in the governance of UPOs as compared to the São Paulo model. The latter was based on a tripartite deliberative council with participation from the real estate sector, academia, social movements and local government. Although formally maintaining the deliberative council, the Rio model symbolized a shift by creating a special purpose company with know-how from real estate and finance aimed at streamlining the management and implementation of the UPO and public-private partnerships in its perimeters. As such, the arrangement further hollows out substantive participation from civil society, a challenge already detected in several UPOs (Pereira, 2016; Salomão, 2016; Maleronka, 2010). Moreover, the contradictory track record of the Rio model (see next section on "perspectives") has shown clear limits to local government's capacity as a "principle" to monitor and evaluate all the actions of its "agents" (i.e. the professional managers and chief executive officers of the special purpose company). In the meantime, the institutional and financial engineering of Rio's *Porto Maravilha* has provided a framework for the design of UPOs in cities such as São Bernardo. The opinions on this financial-institutional arrangement are divided. Contractors, investment managers and financial consultants considered it an emerging model for UPOs. According to one of the contractors we interviewed, the use of specific purpose companies was a sign in the right direction by delegating the design and implementation of UPOs with CEPACs to the private sector. Private sector expertise would both guarantee the development of secondary markets and a sufficiently high and stable flow of earmarked resources to implement the projected infrastructure investments through land value capture. In an indirect critique of the governance model of São Paulo, this was all the public sector-driven management model had not delivered:

> In its present format, UPOs with CEPACS are too sophisticated to be operated by government and too risky for investors.
>
> (Interview no. 7, Contractor, March 2017)

A senior planner in São Paulo was more cautious:

> I am not necessarily against the Rio model provided there is a good project to guide the urban, environmental and social transformation, which is also able to attract private actors. If not, it implies passing on the burden of financial losses in one particular city to Brazilian workers that have contributed to the national low-income housing fund.
>
> (Interview no. 8, São Paulo City Planner 2, December 2016)

4.3.3 Perspectives

Since 2014 Brazil has been facing political and economic turmoil. Large contractors and developers are in jail or being investigated in the context of corruption scandals that have denounced the relationships between private and public corporations and the state. Among the scandals that reached media headlines, the one associated with the UPO *Porto Maravilha* in Rio called particular attention. Initial expectations were structured around a continuous valorization of CEPACs that would generate a financial break-even point of the operation; in 2015 90% of the assets were still unsold, however. The Housing Bank was then pressured to provide additional financial support of R\$ 1.5 billion from the national social housing fund. Ex-congress leader Mr Cunha was accused of having received bribes from developers and contractors involved in Porto Maravilha in order to use his influence over the bank's vice-president (a former fund manager and investment banker), a key person in providing the legitimacy and the additional financial support the drowning port project badly needed.

At the time of writing, low macroeconomic growth, fiscal austerity and large contractors and developers temporarily "out of the game" all suggest that real estate markets and UPOs with CEPACs will go through a period of restructuring. For example, the idea to contract a "market maker" to disseminate this product in the capital market has been postponed by the management of the stock exchange.

At the same time, however, considering the underpriced assets of Brazilian contractors and infrastructure companies, international investment funds such as the Canadian *Brookfield* and the Chinese *State Grid* have looked at the Brazilian crisis as a clear opportunity.[10] According to Fitch Rating senior manager Mauro Storino, "Brazil has become a huge supermarket of assets that is being divided between two players: the Chinese investors with a bag in one hand and Brookfield with one in the other".[11] Both groups have not measured efforts to buy into the depreciated Brazilian energy and infrastructure companies.

Moreover, ongoing regulatory restructuring will further influence the trajectory of UPOs with CEPACs. In addition to earlier-mentioned flexibilization of national housing fund regulation, federal legislation approved in January 2015 (*Statute of the Metropolis*) has now complemented some of the provisions of the CS regarding land-use, urban planning and management at the metropolitan scale. More specifically, the new statute requires that local and state governments collectively discuss and elaborate an integrated development plan at the metropolitan scale. A presidential veto on the creation of a federal fund, however, was a clear sign of the emerging scenario of fiscal austerity, while also not providing any solution for the lack of predictable and transparent sources of finance that has characterized Brazilian metropolitan areas for decades. (Rezende, 2010) As far as finance is concerned, metropolitan areas are expected to work with instruments such as "interfederative" UPOs with CEPACs (by itself an innovation, considering it allows the design and implementation of projects that involve both local and state governments) and public-private partnerships.

4.4 Postscript: Emerging "innovative" financial-institutional devices in time of economic crisis

The logics behind the planning of UPOs with CEPACs is the making of *new* financial assets through the design and implementation of large urban redevelopment projects, driven by successive cycles of increased local government revenues from CEPACs, infrastructure investments and public works undertaken by contractors and valorization of project areas.

As mentioned, considering the key role of contractors, the political crisis of *Lava Jato* left the UPO without some of its main actors, while economic stagnation and the drying up of credit had reduced the attractiveness of large urban redevelopment projects with expensive infrastructure components.[12] Moreover, in this challenging scenario, the financial and legal complexity that accompany the design of any UPOs with CEPAC were also increasingly perceived as a bottleneck. To recall, the City Statute requires public participation and detailed studies on the technical and economic viability, as well as the likely social and environmental impacts of UPOs. Proposals subsequently require approval by the municipal council. In a final stage, the stock exchange evaluates more detailed financial terms of reference, which were drawn up on the basis of the general parameters contained in the municipal law that regulates the UPO.

In 2014, the city administration of São Paulo introduced a new institutional-financial device into its master plan. The so-called PIU (*projeto de intervenção urbana*, meaning "urban intervention project") enabled the city of São Paulo to diversify its planning strategy, until then by and large based on the creation of "new land-based financial assets" through UPO-CEPACs, to one structured around the mobilization of "existing assets" by leasing urban space itself (D'Almeida, 2019).

To be clear, according to the methodology of the master plan that was elaborated and approved during the administration of the Worker's Party (2013–2016), a PIU is a device that was designed in order to establish clearer and more direct linkages between the objective of the master plan and its specific instruments, with an emphasis on the role of large urban redevelopment projects. Thus, the official narrative that was behind the PIUs was that they could improve the design, quality and outreach of large redevelopment projects so as to better contribute to the sustainable transformation of specific areas in the city in line with what was projected in the master plan. As a matter of fact, as was also discussed earlier on, the lack of a clear urban project that guided territorial transformation had been one of the critiques on UPOs with CEPACs, which ended up prioritizing the objective of revenue maximization and, consequently, generating excessively large project areas (Maleronka, 2010).

At the same time, however, the emergence of PIU should be seen in the context of a series of gradual market-enabling transformations in the national institutional framework regarding concessions (Law 8.987/1985) and public-private partnerships (Law 11.079/2004), which had been set in motion during successive federal administrations, including those from the Worker's Party. In that respect, it is worthwhile to mention the rolling out of a national procedure regarding public calls for expression of interest for projects (e.g. decree 8.428 of April 2, 2015), which also mobilized city planners in São Paulo. The specific decree represented an update of earlier legislation, but innovated by including the possibility of proposals from the private sector to trigger a formal public call for expression of interest by the state to receive projects with the objective of generating a subsequent competitive tendering process for concessions and public-private partnerships according to the federal legislation on the subject.[13] This provided an environment whereby contractors and other real estate actors could be involved in the design of projects before the bidding process itself, thereby reducing the risks of getting entangled with badly designed contracts for the implementation of concessions and public-private partnerships.[14]

This combination of market-enabling changes in the national legal framework regarding public-private partnerships, concessions and calls for expressions of interest and the local sense of opportunity generated municipal decree No. 56.901/2016, which fleshed out the specific stages of a PIU. In the first, a diagnosis of the socio-economic and environmental profile of the area as well the guidelines, management structure and pre-viability of a particular redevelopment project should be proposed. This is subsequently discussed with the community and in a third stage, either through a formal call for expression of interest or not, evolves into a more detailed urban redevelopment project (PIU), which requires the definition of the perimeter of areas and the characteristics and successive stages of a project intervention, including the "modelling" of the economic-financial viability and the "democratic management model".[15] This more detailed project proposal is then forwarded for a second round of discussions and deliberations with the community. If approved by the population, the proposal is eventually forwarded for final approval by the municipal council (in case specific parameters of the master plan or zoning law, such as land uses or floor area ratios, are changed) or directly transformed into a redevelopment project by municipal decree (if no changes in relation to the master plan or zoning clauses are implied).

As mentioned, PIUs first appeared as a planning device in the master plan that was approved by the council in 2014. It was subsequently tested, as an emerging praxis, by the Worker's Party administration (2013–2016) through a call for expression of interest to elaborate studies aimed at the ambitious redevelopment project of a derelict industrial corridor along the river Tiete and older railway areas of the city (the *Arco de Tiete* project). This didn't generate enough "raw material" and inputs in order to organize a tender for a specific intervention. To some degree, this was also due to the intensifying crisis that had dramatically affected the financial viability and involvement of contractors in the activities aimed at the design of projects.

Nevertheless, in 2017 the incoming conservative city government was quick to perceive the market-enabling potential of the planning device in terms of articulating public and private actors, planning instruments, programs and financial arrangements around the objective of transforming "existing spaces and infrastructure networks" into financial assets. Although no PIU is actually up and running, the example of how the device is being rolled out by the city government in the concession procedure for twenty-four bus terminals, including the surrounding areas (up to 600 metres), provides a paradigmatic illustration of what is at stake here.[16] To be clear, the 2014 master plan had already zoned these areas around the terminals as "corridors", with the explicit objective of providing incentives for a more intense utilization of the existing transportation networks through densification, higher floor area ratio and permissible mixed-use zoning.[17] Thus, in practical terms, this implied that new projects didn't require any changes in already-permissible land uses, meaning that the earlier-discussed procedure for the collaborative design, discussion and approval of a full-fledged PIU would not need approval by municipal council and could be completed, after the proper discussion with the community, through a relatively fast track municipal decree.

An analysis of the package of twenty-four concessions of bus terminals shows that it is designed around the creative use of the PIU as a device to mobilize the existing land and transportation infrastructure networks, as well the public fund itself as a basis for extracting rents and financial profits. D'Almeida (2019) explores the key ingredients of the approach.

First, and quite different from the previous concession agreements on bus terminals, the PIU package "spatially stretches" the object of the contracts from the terminals to the surrounding perimeter of 600 metres. This means that the cash flow generated by the "residential, service or commercial real estate development are considered accessory income" for investors, interpreted as "part of the broader infrastructure network that is [the] object of the concession agreement" (according to Decree 58.066/2018). This radically inverts the previous municipal legislation, whereby the bidding for concession contracts also included surrounding perimeters but provided for built-in clauses according to which interested parties are required to finance investments and improvements in items such as lighting, access roads and street furniture.[18] Moreover, under the new approach only those assets that are essential to the operation of the bus terminal are required to be transferred back to the local government at the end of the concession period.

Second, the PIU provides a fast-track project-driven approach to the public appropriation of private lands as well as the delivery of public land in the name of the central overriding object of urban redevelopment. In general, expropriation of private land is time-consuming, cumbersome and accompanied by legal court procedures with uncertain outcomes. Nevertheless, according to the interpretation of the municipal attorney general, the PIUs provide a necessary and sufficient basis for the fast-track delivery of public and private land, considering the overriding "public interest" of a project for urban redevelopment (Apparecido Jr, 2017).

Third, and along the same lines, although the Brazilian City Statute provides the general guidelines for what is called an "urban concession" (*concessão urbanistica*),[19] in practical terms the instrument has rarely been used by local governments considering the legal uncertainties and the lack of clarity of what would be the specific object of an urban concession. In other words, while specific infrastructure networks or public works can be framed as objects of concession contracts, it has proven more challenging to design and approve municipal laws on urban concessions. Nevertheless, the municipality of São Paulo did have a legal framework on urban concessions that provided the main financial-legal parameters regarding the redevelopment of areas in partnership with the private sector.[20] On the basis of this legislation and the availability of a general redevelopment project containing objectives, governance structures and financial guidelines, the PIUs that accompany the bus terminals effectively represent a flexible device that provides the justification for land readjustment schemes and urban concessions designed around the generation of rents and financial profits, using both public and expropriated private land within the perimeters of the area. As such, the project that is designed in general terms through the device of a PIU represents more than a concession of specific infrastructure equipment or public works; it provides a concrete perspective for the concession of urban space itself, whereby the specific components of land, infrastructure and public subsidies are mobilized by the private sector in the name of the (local) state. In his analysis on the issue, the municipal general attorney provides a good illustration of this open-ended and flexible interpretation on what constitutes an urban concession (as mobilized by a PIU) (Apparecido Jr, 2017: 257–258):

> This device, by itself, seems to invalidate the argument that the instrument designed through municipal law No. 14.917/09 (*on concessions*) is a simple concession of public works; there is, by the terms authorized by the federal constitution and prevailing federal legislation, a private action, undertaken in the name of the public authority, along the guidelines set out by the latter, structured around urbanistic activities that can be characterized as public works and "non-public works", in addition to the economic responsibilities that neither represent public nor private works or their exploration. The private actor implements *the City Statute and the Master Plan*, the public and private works by themselves representing the main source of renumeration in order to do so, i.e. a means for reaching an end.

4.5 Conclusion

This chapter provided a first case study on the role of financial and non-financial actors alike in the design, implementation and transformation of large urban redevelopment projects. As argued before, this theme has emerged as one of the key elements in the broader research agenda on the role of cities in times of financialization. Our approach was based on the premise that a closer investigation of why, how and by whom specific planning instruments and financial-institutional devices are filled in (or hollowed out) contributes to a better understanding of the limits and potentials of transforming urban space into a tradable income yielding asset. In other words, on the basis of the recognition that urbanism is potentially performative, meaning that its "tool-box" of planning instruments, institutional and financial devices and models not only provide a powerful representation of space but are potentially capable of triggering spatial transformation itself, we have fleshed out the specific experience in the city of São Paulo.

Brazilian Urban redevelopment projects through UPOs with CEPACs represent an example of potential constituent spaces of urban financialization, with an active role for planners within a broader community of financial consultants, investment bankers, builders, contractors and developers, among others, in the physical, financial and contractual design of interventions. Ideally, UPOs linked to CEPACs provide a cost-recovering virtuous cycle, based on successive rounds of additional inflow of revenue through the sale of CEPACs to interested investors from the stock exchange or to users acting in the real estate market, increased investments in infrastructure and (social) housing and valorization of project areas. Really existing UPOs, however, have required significant municipal budgetary funds over and beyond the revenues that were raised through the sales of CEPACs. Secondary markets have not consolidated; as a matter of fact, CEPACs were by and large bought by developers at prices that reflected a circulation of converging norms and expectations regarding the parameters of financial viability studies that were performed in order to maintain the interest of developers and builders. The effective escalation of land and real estate prices, triggered by infrastructure investments, often provided additional challenges to build the low-income housing that was supposed to be part of urban redevelopment projects.

Looking at the specific experience of UPOs with CEPACs, the process of urban financialization still seems rather distant, particularly considering the crisis and restructuring of large Brazilian contractors and ongoing fiscal austerity, which has reduced the perspective of budgetary support that has proven an essential component behind the financial performance of UPOs.

Nevertheless, the recent emergence of more complex arrangements, marked by the combination of UPOs with other instruments such as Public Private Partnerships, real estate investments funds and specific-purpose listed mixed-capital public companies, among others, all suggest an open-ended trajectory driven by variegated strategies aimed at the making of financialization.

In that sense, the recent proliferation of the number and variety of PIUs that has occurred under the present city administration (2017–2020) also points in the same direction. While a first look at the device provides the impression of fragmentation of space through the dissemination of urban redevelopment projects and ad-hoc planning initiatives (which echoes Savini and Aalbers' (2015) description of "de-contextualization of planning" in the specific setting of Sesto San Giovanni, Greater Milan/Italy – see Chapter 1), a closer examination, however, suggests that what is at stake here is the articulation and transformation of existing land, infrastructure networks and public funds around the creation of "abstract financialized space".

Our analysis in this chapter suggests at least four issues for further research. The first is the role of planning instruments and devices in the variegated financialization of urban space. The degree of planning autonomy and "creativity" of Brazilian municipalities provides fertile ground for experimentation with additional building rights, either securitized or not, in the transformation of urban space and finance of housing and infrastructure. While we have prioritized the practical experience in São Paulo, other municipalities have experimented with similar instruments and devices for the design and finance of large urban redevelopment projects. The city of Curitiba, for instance, has used its rather polemic building right quota within its urban policy toolbox of instruments and devices. It is based on the difference between minimum and maximum floor area ratios, as defined and monopolized by the local planning authority. The city has used this device in a discretionary, and rather non-transparent, manner, with little control from the civil society (Da Silva, 2019). During its trajectory, the device has simultaneously been "performed" as a transferable development right (allowing restricted property owners to transfer and sell quota to owners in receiving areas with available infrastructure), a more traditional development impact fee (to be paid to the municipal authority according to pre-established zones and formulas based on the prevailing land values) and as outright credit money. In its latter "monetized" format, the city government has used building right quota in order to pay contractors for the building and reform of infrastructure works such as the World Cup Stadium of 2014. Not unsurprisingly, in relation to the latter example, recurrent cost overruns and inevitable "exchange rate" volatility of the building right quota in light of the economic and real estate crisis have generated financial and legal disputes involving local government, contractors and the owner of the stadium (Da Silva, 2019).

The sharing of risks between the state and private sector in the design and operation of large urban redevelopment projects and contracts for concessions and private public partnerships represents another theme where more analytical and empirical work is needed in order to increase our understanding of the making of urban financialization. Unlike the standard narrative from corporate finance textbooks, which provides a moral link between risk and return, financial investors prefer predictable, contracted and high-income streams (as differential rents) that can capitalize on the present. This is where the state and the planning toolbox come in. Our analysis on UPO-CEPACs and PIUs in the specific context of São Paulo provides ample evidence of how a professional community composed of both financial and non-financial actors alike puts the discounted cash flow models and similar metrics from corporate finance to work through the mobilization of the public fund, infrastructure networks and land in order to provide contracted and predictable income streams. The political economics of valuation, and its implications for the investigation of the calculative practices involved in the "pricing", valuation and sharing of risk in large redevelopment projects, concession and private public partnerships is a field where much more analytical and empirical work is required.

Third, a closer investigation of the entanglements between the public and private sector within emerging governance arrangements also sheds light on the financialization process itself. The overriding narrative that drives ongoing restructuring is the separation between the design and supervision of urban policies (by the state) and the effective implementation of projects and programs in line with these guidelines – undertaken by the professionals from the market that mobilize their financial resources and know-how. The really existing governance arrangements in UPO and PIUs, however, indicate hybridity whereby financial consultants and other intermediaries – either related to developers and contractors or not – increasingly penetrate into the sphere of design as well as supervision of urban policymaking. The proliferation of fast-track public calls for expression of interest in the design and implementation of large urban redevelopment projects with deficient mechanisms for community participation, as well as the dissemination of mixed-capital, specific-purpose companies with significant exposure to the professional know-how, calculative practices and metrics from the financial markets all indicate an increasing pressure to gradually hollow out the collaborative-shared territorial governance that formally underpins the progressive-redistributive project of urban reform and its filling in by premises structured around urban shareholder governance.

Finally, there are multiple spaces of financialization, meaning that the increasing penetration of finance capital into the reproduction of daily life in cities and metropolitan areas is open to contestation. While the international circulation of ideas on the beneficial role of the private sector in terms of allocating the badly needed know-how and financial resources for re-animating our cities has emerged as a hegemonic project, alternative representations of space and finance do exist. In that sense, the work by authors such as Desiree Fields (2014) on the contestation of the contradictory penetration of private equity funds into social rental housing in New York and Berlin, and the elaboration of differential spaces of daily life is emblematic. Along similar lines, Latin American authors such as Maricato (2001) and Rolnik (2018) have produced a critical reflection on the contradictory state mode of production of space, which also inspired and reinforced similar emerging calls for an alternative insurgent praxis on planning and management of cities in the Global South in times of re-emerging global finance.

Notes

1 Section 4.2 and 4.3 are reprinted from the journal, *Land Use Policy, Volume 69*, Jeroen Klink and Lais Eleonora Maróstica Stroher, "The Making of Urban Financialization? An exploration of Brazilian urban partnership operations with building certificates", pages 521–526 (2017), with permission from Elsevier.

2 CEPAC means *Certificado de Potencial Adicional de Construção* (Additional Building Right Certificate).

3 In the words of one of the brokers we interviewed: "I think this mechanism of securitized building rights is wonderful business; it represents an emerging Brazil, bringing in the awareness and responsibilities regarding capital markets" (Interview no. 1, Broker, November 2016). Citation taken from Klink and Stroher (2017: 519).

4 Maleronka (2010: 129) analyzes that while *Água Espraiada* incorporated 1,400 hectares, the ZAC *Paris Rive Gauche* and Poblenau in Barcelona comprised 130 hectares.
5 The Curitiba scenarios were elaborated on the basis of 10% cost overruns.
6 Explanations for the agio in 2008 remain inconclusive, such as buyer errors predicting finalization of the UPO and individual efforts to speculate with CEPACs.
7 These values were also converted to 2013.
8 For Faria Lima, for example, the initial emission on 26/5/2010 provided a price of R\$ 6,343.47, while on 15/10/2015 CEPACs were sold in the secondary market for R\$ 11,083.23. (Data organized by the authors on the basis of prospectuses and interviews with CEPAC brokers.)
9 More specifically, national decree No 13, May the 10th of 2016 (Ministry for Cities).
10 Brookfield has signed a preliminary agreement to buy Odebrecht Ambiental for 1.7 billion USD.
11 Fusões e Aquisições. *Brookfield avança com Lava Jato e 'pechinchas'*. Available at: http://fusoesaquisicoes.blogspot.com.br/2017/01/brookfield-avanca-com-lava-jato-e.html. (Accessed 16 January 2017).
12 *Lava Jato* -Car Wash- is the name of the investigation undertaken by the National Public Prosecutor's Office and federal judges of possible bribes and corruption involving state companies such as Petrobras (active in oil, gas and refinery), development banks and contractors of public works. Among others, it generated the imprisonment of former president *Lula Inácio da Silva*, as well as the owners and several chief executive officers of contractor companies and state enterprises.
13 The subsequent competitive bidding not only allows the original firms to participate but also envisages the possibility of compensation of the costs associated with the initial study.
14 Contractors were keen on increasing their leverage over the design of redevelopment projects in general, and risk sharing clauses in particular. See, for instance, the introduction to the project proposal that was developed for the Arco de Tiête redevelopment project by the contractor Odebrecht: "Defining the risk regarding costs and timing of the implementation of works as a responsibility of the private partner, the latter should be in the position to decide on the commercialization of building rights. To do the opposite would be to transform risk (probable event) into uncertainty (unpredictable event), putting private partners into a situation where they have no control" (Odebrecht et al., 2016: 2–3). See D'Almeida (2019) for a more detailed discussion on the Arco de Tiête project in the city administration of former mayor Fernando Haddad.
15 According to municipal Decree No 56.901/2016, providing guidelines for Urban Intervention Projects (PIUs).
16 The following discussion is based on D'Almeida (2019).
17 The plan allowed for floor area ratios of four (that is, building rights four times the size of the land area), accompanied by generous bonuses (additional building rights) in case developers followed certain municipal guidelines in terms of the use of façades and circulation area for pedestrians.
18 To be specific, municipal law 16.211/2015 was changed to law 16.703/2017.
19 The City Statute provides general guidelines on what is called "concerted urbanism", that is, "the cooperation between governments, private sector and Civil Society in the urbanization process, while attending to the social interest".
20 More particularly, municipal Law 14917/09, approved on May 7, 2009, regulates urbanistic concessions in the municipality of São Paulo.

References

Apparecido Jr JA (2017) *Direito urbanístico aplicado: os caminhos da eficiência jurídica nos projetos urbanísticos*. São Paulo: Juruá.

Blanco AGB; Cibils VF and Muñoz AFM (2016) *Expandiendo el uso de la valorizacion del suelo*. Washington: IADB.

Cardoso IC (2013) O papel da Operação Urbana Consorciada do Porto do Rio de Janeiro na estruturação do espaço urbano: uma "máquina de crescimento urbano"? *O Social em Questão* 16(29): 69–100.

Centro de Estudos da Metrópole (Cem). *Lançamentos Imobiliários Comerciais e Residenciais da Base da Embraesp de 1985 a 2013*. Available at: www.fflch.usp.br/centroda metropole/v3/bases.php?retorno=716&language=pt_br (Accessed 20 December 2016).

Coma MC (2011) Del sueño olímpico al proyecto Porto Maravilha: el "eventismo" como catalizador de la regeneración a través de grandes proyectos urbanos. *Revista Brasileira de Gestão Urbana* 3(2): 211–227.

D'Almeida CH (2019) *Concessa Venia. Estado, empresa e a concessão da Produção do Espaço Urbano*. Tese de Doutorado. Instituto de Arquitetura e Urbanismo, Universidade de São Paulo, São Carlos.

Da Silva M (2019) *O fetiche dos instrumentos de solo criado: A experiência da aplicação da Cota de Potencial Construtivo em Curitiba*. Dissertação de Mestrado. Faculdade de Arquitetura e Urbanismo, Universidade de São Paulo, São Paulo.

DiPasquale D and Wheaton W (1996) *Urban economics and real estate markets*. Upper Saddle River, NJ: Prentice Hall.

Fields D (2014) Contesting the financialization of urban space: Community organizations and the struggle to preserve affordable rental housing in New York City. *Journal of Urban Affairs* 37(2): 144–165.

Fix M (2009) Uma ponte para a especulação – ou a arte da renda na montagem de uma "cidade global". *Caderno CRH* 22(55): 41–64.

Halbert L and Attuyer K (2016) Introduction: The financialization of urban production: Conditions, mediations and transformations. *Urban Studies* 53(7): 1347–1361.

Klink J and Stroher LEM (2017) The making of urban financialization? An exploration of urban partnership operations with building certificates. *Land Use Policy* 69: 519–528.

Maleronka C (2010) *Projeto e Gestão na Metrópole Contemporânea: um estudo sobre as potencialidades do instrumento operação urbana consorciada à luz da experiência paulistana*. Tese de Doutorado. Faculdade de Arquitetura e Urbanismo, Universidade de São Paulo, São Paulo.

Maricato E (2001) *Brasil, Cidades. Alternativas para a crise urbana*. Petrópolis: Editora Vozes.

McAllister P (2017) The calculative turn in land value capture: Lessons from the English planning system. *Land Use Policy* 63: 122–129.

Odebrecht; OAS and URBEM (2016) *I Pilares do projeto. II Empresa SA Complementacão à 2O fase do PMI do Arco Tiête*. São Paulo: Prefeitura de São Paulo (Documento).

Ondetti G (2016) The social function of property, land rights and social welfare in Brazil. *Land Use Policy* 50: 29–37.

Pereira A (2016) *Intervenções em centros urbanos e conflitos distributivos: modelos regulatórios, circuitos de valorização, e estratégias discursivas*. São Paulo: Universidade de São Paulo.

PMC (Prefeitura Municipal de Curitiba) (2012) *Prospecto da Operação Urbana Consorciada Linha Verde*. Curitiba: PMC.

PMC (Prefeitura Municipal de Curitiba) (2015) *OUC Linha Verde. Relatório Trimestral CVM*. Curitiba: PMC.

PMRJ (Prefeitura Municipal do Rio de Janeiro) (2012) *Prospecto da Operação Urbana Consorciada do Porto do Rio de Janeiro*. Rio de Janeiro: PMRJ.

PMSBC (Prefeitura Municipal de São Bernardo do Campo) (2015) *Estudo de Impacto de Vizinhança (EIV)*. *Operação Urbana Consorciada São Bernardo do Campo*. São Bernardo do Campo: PMSBC.

PMSP (Prefeitura Municipal de São Paulo) (2008) *Prospecto da Operação Urbana Consorciada "Espraiada"*. São Paulo: PMSP.

PMSP (Prefeitura Municipal de São Paulo) (2012) *Prospecto da Operação Urbana Consorciada "Faria Lima"*. São Paulo: PMSP.

PMSP (Prefeitura Municipal de São Paulo) (2014a) *Prospecto da Operação Urbana Consorciada "Branca"*. São Paulo: PMSP.

PMSP (Prefeitura Municipal de São Paulo) (2014b) *Edital do Primeiro Leilão da Distribuição pública de CEPAC da Operação Urbana Consorciada "Branca"*. São Paulo: PMSP.

PMSP/SP Urbanismo (2015a) *Histórico de Leilões. OUC Espraiada*. São Paulo: PMSP.

PMSP/SP Urbanismo (2015b) *Histórico de Leilões. OUC Faria Lima*. São Paulo: PMSP.

Rezende F (2010) Em busca de um novo modelo de financiamento metropolitano. In: Magalhães F (ed.) *Regiões Metropolitanas no Brasil. Um paradoxo de desafios e oportunidades*. Washington: Banco Interamericano de Desenvolvimento, 264p.

Rolnik R (2018) *Urban warfare: Housing under the empire of finance*. London: Verso.

Royer L (2014) *Financeirização da Política Habitacional: Limites e Perspectivas*. Vol. 1. São Paulo: Annablume.

Salomão TMN (2016) *Linguagem técnica e (im)possibilidade para a produção democrática do espaço urbano: uma análise a partir de duas experiências participativas em Belo Horizonte*. Belo Horizonte: UFMG.

Sandroni P (2010) A new financial instrument of value capture in São Paulo: Certificates of additional construction potential. In: Ingram GK and Hong YH (eds.) *Municipal revenues and land policy*. Cambridge, MA: Lincoln Institute of Land Policy.

Savini F and Aalbers MB (2015) The de-contextualization of land use planning through financialization: Urban redevelopment in Milan. *European Urban and Regional Development Studies*: 1–17.

Siqueira MT (2014) Entre o fundamental e o contingente: dimensões da gentrificação contemporânea nas operações urbanas em São Paulo. *Cadernos Metrópole*, São Paulo 16(32): 391–415.

Smolka M (2013) *Implementing value capture in Latin America: Policies and tools for urban development*. Cambridge: Lincoln Institute of Land Policy.

Weber R (2010) Selling city futures: The financialization of urban redevelopment policy. *Economic Geography* 86(3): 251–274.

5 On contested water governance and the making of urban financialization

Exploring the case of metropolitan São Paulo, Brazil[1]

5.1 Introduction

This chapter provides a second case study that illustrates the potential of a political economy of valuation for the investigation of the making of financialization in cities of the Global South. It is aimed at filling in some of the gaps in the literature along the lines discussed in Chapter 2 through a case study on the gradual transformation of a shared state-municipal arrangement for metropolitan water governance into a shareholder-driven system in São Paulo, Brazil. More specifically, since the early 2000s, the mixed-capital, majority state-owned water and sewage company São Paulo Company for Basic Sanitation (SABESP) (listed on the São Paulo and New York stock exchange) has been able to provide "value for money" to its shareholders through a consistent increase in profitability and share prices.

More specifically, the chapter shows that its aggressive calculative practices used in the pricing of wholesale water delivered to municipal retail companies, which were homologated by regulatory agencies, legal courts and the federal antitrust body, were key in the accumulation of municipal debts with SABESP. Considering the gradual consolidation of legal impasses, whereby debts should be paid one way or another, this created an "enabling" environment for the negotiation of municipal debt relief in exchange for awarding SABESP the concession to build, operate and maintain municipal networks as well as to design and implement pricing and revenues strategies. On the basis of a detailed analysis of the disputes between municipal utility Environmental Sanitation Services of Santo André (SEMASA), owned by the city of Santo André (approximately 720.000 inhabitants, located in metropolitan São Paulo) and SABESP since the mid-1990s, we show the role of financial-legal devices and specific forward- and backward-looking calculative practices in the pricing of water and valuation of assets, and the subsequent penetration of shareholder value premises into a shared state-municipal arrangement for metropolitan water governance.

Beyond the insights from the case, indicating that financialization of water in metropolitan São Paulo is still open-ended, this chapter provides two contributions to debates on urban financialization with relevance for the Global South.

First, different from the US-based work on austerity that has led cities to adopt increasingly risky strategies in capital markets (Kirkpatrick, 2016; Peck and

Whiteside, 2016), as well the literature on the entanglements between finance, real estate, large urban development projects and governance in the European context (Van Loon et al., 2018), our analysis provides insights regarding how, in a context of relatively thin capital markets, intergovernmental conflicts between state and municipalities regarding the regulation, valuation and pricing of water are instrumental in understanding the "making of urban financialization". We show how the circulation of particular calculative practices that involve legal courts, regulatory agencies, consultancies and public utilities enable the accumulation of intergovernmental debt, which contributes to the gradual hollowing out of shared governance and its filling in with shareholder water governance.

Second, while political economy has provided an understanding of the broader contradictions associated with urban neoliberalization, entrepreneurialism and the penetration of finance capital into cities by transforming them into "tradable income yielding assets" (Guironnet and Halbert, 2015; Lapavitsas, 2013; Weber, 2010; Zwan, 2014), it remains unclear how this has occurred (Callon, 1998; Hall, 2010; MacKenzie, 2006). More specifically, urban political economy has prioritized the analysis of cities as privileged arenas for the generation and extraction of financial profit, while taking markets relatively for granted. As such, it has traditionally spent less effort in understanding the constitution of markets as key arenas for the circulation and consumption of value, prioritized by social studies of finance (Christophers, 2014). Considering the technological indivisibilities, significant state presence in the planning of housing, land and infrastructure networks, long payback periods and political contestations that surround this process, there is nothing inherent in the transformation of cities into financial assets. Thus, following Christophers' (2014) general argument, we provide a first illustration of how a more articulate approach between political economy and social studies of finance might contribute to understanding the making of financialization in cities and infrastructure networks.

This case study in metropolitan São Paulo was developed through documentary research on SABESP and SEMASA, a detailed analysis of the files of SEMASA's request to investigate SABESP, which were obtained from the national anti-trust agency CADE, and complementary secondary data collected for metropolitan São Paulo.

Three sections follow this introduction. The first summarizes the institutional trajectory of sanitation in Brazilian metropolitan regions, while the second fleshes out the financial-institutional battlefield between SABESP and SEMASA. We conclude wrapping up the case with suggestions for further research on urban financialization.

5.2 The institutional trajectory of sanitation in Brazilian metropolitan areas

It is not our purpose to analyze in detail the institutional setup of sanitation in Brazilian metropolitan areas.[2] Table 5.1, based on Aversa (2016) and Barcellos de Souza (2013), provides an overview of the successive stages of the Brazilian

Table 5.1 Governance of basic sanitation in Brazilian metropolitan regions

Periodization	1875–1915–30	1915–30–1964	1964–mid-1980s	mid-1980s–mid-2000s	mid-2000s–2015
Geographical and historical development trajectory	Predominantly agro-export-oriented regime.	Emerging metropolitanization & rise of national developmentalism.	Intensification of metropolitanization & authoritarian national developmentalism	Neoliberalization, restructuring, rescaling and roll-back of national developmental regime.	Social/neo-developmental regime, recently followed by renewed round of rolling back of the state.
Governance Regime in sanitation	Private	Shared techno-bureaucratic (local & state governments).	Centralized techno-bureaucratic (federal finance; state planning and management; incentives aimed at transfer of municipal utilities to state).	Economic crisis, austerity and hollowing out of techno-bureaucratic regime. Rise of (neo)private governance.	National system for shared collaborative governance (state, municipal, civil society), contested by renewed (neo)private/corporate and financialized arrangements.
Legal frameworks in sanitation	Inexistent	Collaborative arrangements between state and local governments: (e.g. São Paulo: role of CODEGRAN, COMASP and SANESP: state and municipalities share ownership of metropolitan specific-purpose public utilities).	PLANASA (National Water Supply and Sanitation Plan) and National Housing Bank (NHB); (e.g. São Paulo: creation of State company SABESP (1973), aimed at centralization of planning, management and finance. Emerging conflicts with municipal utilities).	Bankruptcy NHB and demise of PLANASA. Institutional vacuum regarding basic sanitation.	Institutional strengthening of regime aimed at shared governance, while maintaining insecurities: National framework law on basic sanitation (11.445/2007); Constitutional jurisprudence (2013) on shared responsibilities (states and municipalities) for sanitation in metropolitan areas (ADI – 1.842).
Metropolitan Governance regime	Inexistent	By and large inexistent; *CODEGRAN* in metropolitan São Paulo.	Centralized at federal level; authoritarian and techno-bureaucratic regime.	Post 1988 constitutional vacuum regarding metropolitan governance.	Institutional strengthening through national framework laws (e.g.: Statute of the Metropolis – 13.089/2015). Persistent deficiency in fiscal federalism for metropolitan governance.

Source: Author's elaboration. Based on Aversa (2016: 8, 14) and Barcellos de Souza (2013).

developmental state restructuring and its relations with metropolitan governance and sanitation.

While developmental policies and industrialization had already been set in motion since the 1930s, being accompanied by a rural-urban transition and growth of big cities, it was not until the military regime, which took over democratic rule in 1964, that a highly centralized technocratic system of governance and finance was established for Brazilian metropolitan areas, with a key role for institutions like the National Housing Bank (NHB) and the national system for sanitation (PLANASA). Metropolitan finance has historically lacked a system of predictable, re-distributive federal transfers of tax resources to smoothen intra-metropolitan disparities (Rezende, 2010).[3] Funding for NHB and PLANASA was based on a compulsory surcharge on wages that was linked to an unemployment and retirement fund. As such, the guiding principle behind funding was (partial) cost recovery of investments that were mainly directed to social housing and basic sanitation. Federal government was keen on stimulating, through selective financial incentives to cities, the voluntary transfer of municipal utilities to the state sanitation companies that were being created. In São Paulo, for example, this triggered large-scale transfers of municipal utilities to state company SABESP, which had been created in 1973. The approach also signalled the hollowing out of the prevailing model of shared governance and a gradual buildup of tensions between SABESP and municipalities that decided to maintain their water company.

From the mid-1980s onwards, this "peripheral" developmental infrastructural regime (considering that, unlike welfare regimes prevailing in most industrialized countries, it had never channelled federal tax resources to subsidize and connect poor target groups to the city and its networks) was collapsing in light of macroeconomic stagnation, hyperinflation and the bankruptcy of the NHB and its financial model. Moreover, the disarray of developmental institutions increased the pressure to privatize state sanitation companies (Britto and Rezende, 2017). Although SABESP was not privatized and maintained majority state ownership, it was restructured and got listed on the São Paulo (1996) and NY stock exchange (2002).

After 2005, the scenario indicates institutional strengthening aimed at shared governance involving state, municipalities and civil society. The national framework legislation on sanitation of 2007 was key in that it provided guidelines for the planning (municipalities), operation and management (municipal or state concessionaries) and regulatory supervision of services, which were to be created by granting authorities. Likewise, a 2013 supreme court ruling (ADI 1.842) confirmed that neither state nor local governments have exclusive responsibility over sanitation in metropolitan areas, but they need to design specific collaborative governance arrangements.[4] Nevertheless, this tendency towards collaborative governance has not deterred the proliferation of institutional and financial conflicts between municipalities and states, as will be illustrated subsequently.

5.3 The financial-institutional battlefield in Greater São Paulo[5]

5.3.1 *From shared to shareholder governance?*

SABESP currently supplies 364 of the state's 645 municipalities. Around 70% of its gross revenue from water and sewage is obtained from the 39 cities that make up metropolitan São Paulo, which represents both a key market and a risk factor within its business strategy. The latter is understood when considering that the "supreme court still has not clarified the effects and the extension of its decision ADI 1842 in 2013" (SABESP, 2017: 20) (i.e. regarding how the shared responsibility of states and municipalities in metropolitan areas should be filled in). Moreover, several metropolitan cities had maintained their utilities since 1973, while others that had granted SABESP a concession to build/invest, operate and maintain were now approaching the stage to decide on renewal with SABESP, or re-assume their municipal networks.

The concession negotiation with the city of São Paulo, which provides around half of its gross operational revenue, illustrates well SABESP's metropolitan strategy. Conversations started immediately after the creation of the regulatory agency, Regulatory Agency for Sanitation and Energy of the State of São Paulo (ARSESP) in 2007, along the framework of national law No 11.445/2007. ARSESP is responsible for the regulation and monitoring of sanitation services (including the provision of binding guidelines for pricing), which are either under SABESP's direct control or where it acts as a concessionaire of local governments. It also performs a key role in the resolution of disputes in concession contracts. As part of the executive structure of the state of São Paulo, and receiving its support in terms of staffing and finance, its independency has been questioned. Rating agency Moody (2013: 2), for example, considered that:

> The short period of operation of ARSESP and the lack of strong evidence of complete independence from the state government impose additional regulatory risks. The board of Directors is composed of members appointed by the state government and approved by parliament.

The city of São Paulo was pressured to grant a 30-year concession in light of the significant debt it had accumulated with SABESP. The contract that was signed in 2010 provided SABESP with the exclusive right to invest/build, operate and maintain the network, irrespective of the specific design for shared governance that would be adopted according to the 2013 supreme court ruling. Moreover, revenue sharing clauses with the city of São Paulo were conditioned to financial viability of the contract, while rules regarding contractual suspension (requiring inflation-adjusted payment of non-depreciated assets plus a 15% surcharge) provided additional guarantees for SABESP (Ferreira, 2019).

Figure 5.1 illustrates the evolution of SABESP's nominal share prices since its listing on the NY stock exchange in 2002 until its transformation in a public

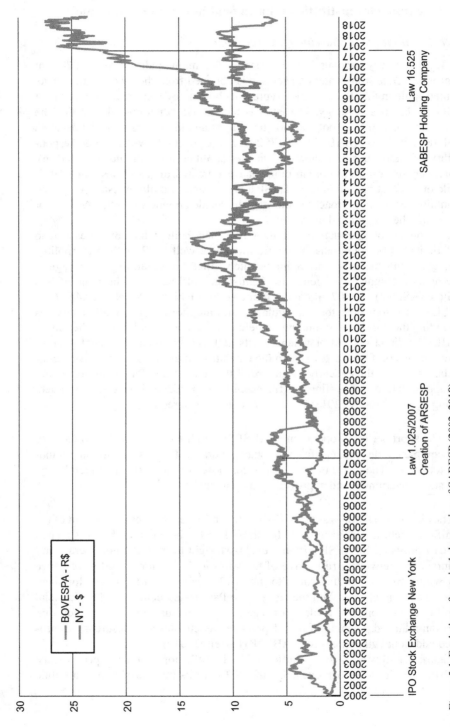

Figure 5.1 Evolution of nominal share price of SABESP (2002–2018)

Source: Ferreira, 2019.

holding company in 2017.[6] The performance on the São Paulo exchange shows a yearly increase of almost 200%, or 65% in real prices of 2002.[7] With the exception of the drought and water crisis during 2013–2015, shares have been rising since the creation of ARSESP in 2007. Nevertheless, part of SABESP's financial surplus leaked away; during 2007–2016, the company generated a net profit of R$ 14.6 billion, of which R$ 10.3 was reinvested.[8] As a majority shareowner, the state received R$ 2.16 billion (out of R$ 4.3) in dividends, which were allocated externally to finance activities out of the general budget (Henrique, 2017: 124).

5.3.2 Institutional conflicts in the metropolitan fringe

SABESP's strategy intensified conflicts in other metropolitan cities such as Santo André, which also hadn't granted a concession. In the 1990s, its utility SEMASA was the first Brazilian company to integrate solid waste, water, sewage and drainage within an integrated urban development perspective.

In 2015 SEMASA filed a request to the National Antitrust Agency (the Administrative Council for the Protection of the Economic Order (CADE)) to investigate infringements of the economic order committed by SABESP. SEMASA claimed that, since the mid-1990s, SABESP had been using its quasi-monopoly position in the wholesale market in order to apply abusive, margin-squeezing pricing policies to municipal utilities, which depended on SABESP's supply to provide retail water to households.[9] Using international jurisprudence on the "as efficient competitor" test used in similar situations,[10] it calculated that SABESP would be unable to maintain viability of its own retail operations under the prevailing discriminatory wholesale prices it practiced. It estimated that SABESP would accumulate losses in its retail operations around R$ 59 to R$ 80 billion during 1999 and 2014, considering the prices it charged to SEMASA.

SEMASA showed that the disagreements regarding water pricing had generated a non-justified escalation of municipal debt with SABESP, which is illustrated in Table 5.2 (in present values of December 2014).[11] Table 5.2A shows the volume of water and sewage delivered by SABESP (in 1000 m³), the average yearly tariff it effectively charged as compared to the lower tariff associated with the "as efficient competitor tariff". Table 5.2B subsequently shows the unjustified accumulated stock of debt, that is, the difference between tariff revenue charged by SABESP and the amount of revenues associated with the lower tariff. The interest rates charged over outstanding debts in similar court cases involving SABESP are summarized in Table 5.2C, providing an inflation-adjusted estimate of total accumulated debt (amortization plus interest), during 1999–2014, of around R$ 1.4 billion (Table 5.2D).

Similar disputes had occurred in other cities in the metropolitan fringe, such as Osasco (1999), São Bernardo do Campo (2003) and Diadema (2013), and were instrumental in SABESP's strategy to eventually control the metropolitan retail market for water. In all these cases, negotiations outside the courts regarding debt relief in exchange for concessions awarded to SABESP had been completed and signed. According to SEMASA, SABESP's excessive wholesale prices had

Table 5.2 Pricing and intergovernmental debt: SEMASA's contestation (*)(**)

A) Flows: Volumes sold (1000m3) and Tariffs (in R$/1000m3)

	C.1	C.2	C.3
	Volume sold	Tariff Effectively charged by SABESP	As efficient competitor tariff calculated by SEMASA
1999	57,106	R$1,218.31	R$499.00
2000	55,821	R$1,253.11	R$518.87
2001	52,319	R$1,258.38	R$546.54
2002	54,827	R$1,261.62	R$558.36
2003	58,283	R$1,282.74	R$535.85
2004	57,111	R$1,379.33	R$608.67
2005	58,189	R$1,413.63	R$583.82
2006	58,681	R$1,501.15	R$629.90
2007	64,549	R$1,518.13	R$614.19
2008	65,743	R$1,482.09	R$605.30
2009	66,169	R$1,499.96	R$605.40
2010	70,277	R$1,476.77	R$560.18
2011	71,195	R$1,464.72	R$606.51
2012	71,204	R$1,483.71	R$580.80
2013	73,749	R$1,503.37	R$574.81

B) Stocks: Revenues charged by SABESP vs Revenues according to the "as efficient competitor" tariff

	C.1	C.2	C.3
	Revenue due to SABESP (C.1 X C.2, Table 2A)	Revenue due with as efficient competitor tariff (C.1 X C.3, Table 2A)	Accumulated debt levels (C.3 = C.1–C.2)
1999	R$69,573,000	R$28,495,894	R$41,077,106
2000	R$69,950,000	R$28,963,842	R$40,986,158
2001	R$65,837,000	R$28,594,426	R$37,242,574
2002	R$69,171,000	R$30,613,204	R$38,557,796
2003	R$74,762,000	R$31,230,946	R$43,531,054
2004	R$78,775,000	R$34,761,752	R$44,013,248
2005	R$82,258,000	R$33,971,902	R$48,286,098
2006	R$88,089,000	R$36,963,162	R$51,125,838
2007	R$97,994,000	R$39,645,350	R$58,348,650
2008	R$97,437,000	R$39,794,238	R$57,642,762
2009	R$99,251,000	R$40,058,713	R$59,192,287
2010	R$103,783,000	R$39,367,770	R$64,415,230
2011	R$104,281,000	R$43,180,479	R$61,100,521
2012	R$105,646,000	R$41,355,283	R$64,290,717
2013	R$110,872,000	R$42,391,663	R$68,480,337

C) *Interest charged over contested stock of debts*

	C.1	C.2	C.3
	Interest (%) (***)	*Accumulated debt level (Table 2B , C.3)*	*Interest payments due (C.3 = C.1 X C.2)*
1999	156%	R$41,077,106	R$64,080,285
2000	150%	R$40,986,158	R$61,479,237
2001	144%	R$37,242,574	R$53,629,306
2002	138%	R$38,557,796	R$53,209,759
2003	132%	R$43,531,054	R$57,460,992
2004	120%	R$44,013,248	R$52,815,897
2005	108%	R$48,286,098	R$52,148,986
2006	96%	R$51,125,838	R$49,080,805
2007	84%	R$58,348,650	R$49,012,866
2008	72%	R$57,642,762	R$41,502,789
2009	60%	R$59,192,287	R$35,515,372
2010	48%	R$64,415,230	R$30,919,310
2011	36%	R$61,100,521	R$21,996,187
2012	24%	R$64,290,717	R$15,429,772
2013	12%	R$68,480,337	R$8,217,640
		R$778,290,376	**R$646,499,204**

D) *Consolidated Non-justified Accumulation of Debt*

	C.1	C.2	C.3
Index	*Principal (Table 2C, C.2)*	*Interest (Table 2C, C.3)*	*Total (C.3 = C.1 + C.2)*
CPI/FIPE	**R$778,290,376**	**R$646,499,204**	**R$1,424,789,580**

Source: Author's elaboration. Based on CADE, 2018.

(*) In Present values of December 2014. Consumer Price Index FIPE.
(**) R$ 3,66= 1 USD (as per 02/02/2019)
(***) Until 2003 the yearly interest rate charged was 6%; after 2003 this increased to 12%.

squeezed financial margins of all these municipal utilities, considering the political challenge to pass on these costs to poor households through escalating retail prices. In the meantime, the escalating stock of debt of metropolitan cities Guarulhos, Mauá and Santo André, all objects of pending court cases and parallel negotiations with SABESP, was estimated around R$ 8 billion (present values of March 2016). SEMASA would be the next in a series of "forced" signatures of concession contracts with SABESP, which would further hollow out shared water governance in metropolitan São Paulo.

SEMASA also questioned state regulatory agency ARSESP; while created in 2007, it only managed to publish its first four-year guidelines on the criteria for water tariffs in 2014 (see next section). Until then, ARSESP had never worried about evaluating SABESP's pricing rules and simply approved yearly inflation adjustments of pre-established tariffs.

In 2017, CADE rejected SEMASA's request to investigate SABESP on three arguments.[12] First, although it had a quasi-monopoly position in the wholesale water market, SABESP had no dominance in retail. Rather, in metropolitan cities with municipal utilities, such as Santo André, this monopoly was exercised locally. Second, CADE argued that price revisions had occurred within guidelines set and monitored by an independent regulatory agency (ARSESP), which also ruled out possible advantages of discriminatory pricing in wholesale and retail markets. Finally, while stating it was not its responsibility to evaluate specific pricing rules adopted by two market players, CADE nevertheless argued that SEMASA's request was seen in the context of the latter's refusal to raise retail water prices – which could have avoided its debt – rather than an infringement of the economic order by SABESP. Being a local monopolist, CADE considered SEMASA could have easily raised its retail prices (CADE, 2018: 12):[13]

> Especially considering a downstream and heavily indebted monopolist, its pricing rule should consider passing on to consumers the rising costs of inputs. Instead, what seems to be happening in this case is that SEMASA reduces its costs by paying less than it should to SABESP with the objective of not charging higher tariffs to the inhabitants of Santo André.

5.3.3 What price for water?

Understanding what is at stake requires an analysis of the logic behind the water tariff revisions coordinated by ARSESP, which are binding for SABESP and cities where it performs the role of concessionaire.

ARSESP adopts a price-cap rule, setting a ceiling for tariffs that guarantees a minimum rate of return on investments and operating expenditures and financial viability for debt and equity providers. This price ceiling should also allow the expansion of service coverage and quality goals as targeted by municipal and state sanitation plans, while avoiding excessive burdens for consumers. The formula used is based on the financial literature on long-run average incremental cost pricing for infrastructure (Bahl and Linn, 1992). The technique applies forward-looking discounted cash flow (DCF) analysis, providing a water tariff that breaks even, in net present values terms, the projected costs and revenues over the life-cycle of projects, while using a minimum required rate of return equivalent to the weighted average, risk-adjusted cost of capital for debt and equity finance. In the specific case of SABESP, the discount rate used during the first tariff cycle (2011–2014) was 8,06%.

Nevertheless, different from pure, forward-looking DCF methods (capitalizing expected cash flows of revenues and expenditures), ARSESP also includes historical cost accounting to evaluate SABESP's non-depreciated engineering objects (networks, pipelines, buildings and so on). Therefore, in a variation on Chiapello's (2015) analysis on the substitution of historical cost methods by forward-looking financial DCF analysis in valuation, ARSESP adopts a "hybrid" approach, whereby historical cost accounting of non-depreciated investments undertaken in the past and expected financial cash flows in the future are combined. In a way, non-depreciated investments undertaken by SABESP in the past, as engineering objects, become assets once they are incorporated into ARSESP's "regulatory baseline", considering these also receive the required minimum rate of return.

The concrete perspective of assetization (Birch, 2017) (i.e. in this particular case, artificially inflating the accounting base of non-depreciated investment of the past) created a series of conflicts regarding the regulatory process in general and the way it could facilitate price increases charged to municipal retailers such as SEMASA, in particular. On the one hand, SABESP argued, and anti-trust agency CADE agreed, that charging more in retail markets would imply lower prices in the wholesale segment (and vice versa), considering that the price-cap regime set a ceiling for the rate of return and effectively consolidated a zero-sum game for discriminatory pricing strategies. On the other, SEMASA contested that CADE had clearly underestimated the possibility of creative calculative practices aimed at increasing ARSESP's regulatory baseline for SABESP, which would enable the transformation of engineering objects into assets by inflating non-depreciated investments undertaken in the past.

Table 5.3 provides a three-year simplified, hypothetical illustration of SEMA-SA's argument (investments occurring in the first, net operational revenues in subsequent two years.) Base scenario A shows a pure DCF approach making decisions in year 1. Expected investment costs of R$ 100 billion (representing a yearly accounting depreciation cost of R$ 50 billion in the loss-profit statement) require a projected average annual revenue of R$ 57.21 billion in order to recover a hypothetical cost of capital of 10% (resulting in a zero net present value of the investment). Scenario B illustrates hybrid DCF accounting where revaluation of non-amortized investments of the past penetrates into calculations for the future. Assuming the project is "up and running" at the end of the second year, what happens if the value of non-depreciated investments of the past (year 1), which can be interpreted as ARSESP's regulatory asset base, are inflated from R$ 50 billion in the base scenario (R$ 100 billion–R $50 billion) to R$ 70 billion (R$ 120 billion–R $50 billion)? The figures in bold-italics of scenario B show that this requires increasing expected cash flow revenues for the third year to R$ 81.82 billion (being the sum of the initial projection for the third year, i.e. R$ 57.21 billion, and the increase in the regulatory asset base (R$ 20 billion), adjusted with the compounded cost of capital of 10% during two years).

Thus, even with a cap on the overall rate of return in wholesale and retail markets (being 8,06% for SABESP), SEMASA claimed that in a context of cumbersome access to SABESP accounts,[14] uncertainty and complexity (networked metropolitan systems being amortized over decades), malpractice of the data

Table 5.3 Regulatory base and calculative practices: A hypothetical illustration

Assumptions:				
Weighted Average Cost of Capital (wacc)	10%			
Projected Investment cost in base scenario A	R$ 100			
"Inflated" Regulatory Base in scenario B	R$ 120			

A. Base scenario – Decision-making in year 1

Pure Discounted Cash Flow Method (Future-oriented, decision-making start of year 1)

Year	NPV (Year 1)	1	2	3
(1) Expected net operational revenues			R$57.62	R$57.62
(2) Projected investment costs		R$100.00		
Net Cash Flow (= 1–2)	R$0.00	–R$100.00	R$57.62	R$57.62
	(0 = –100+ 57.62/1.1+ 57.62/1.21)			

Loss-Profit Statement for year 2 and 3 (Associated with Investment decision made in year 1)

Year	NPV (Year 1)	1	2	3
(1) Realized net operational revenues			R$57.62	R$57.62
(2) Depreciation			R$50.00	R$50.00
(3) Interest (wacc) on non-depreciated capital			R$10.00	R$5.00
Net Profit (= 1–2–3)	R$0.00	R$0.00	–R$2.38	R$2.62
	(0= 0–2.38/1.1+2.62/1.21)			

B. Scenario with Inflated regulatory base – Adjusting the numbers in year 2 (for year 1 & 3)

"Hybrid" Discounted Cash Flow Method ("Adjusting the past and the future")

Year	NPV (Year 1)	1	2	3
(1) Adjusted Expected net operational revenues			R$57.62	**R$81.82**
(2) "Inflated" Investment Base (assetization)		**R$120.00**		
Net Cash Flow (= 1–2)	R$0.00	–R$120.00	R$57.62	R$81.82
	(0= –120+ 57.62/1.1+ 81.82/1.21)			

Adjusted Loss-Profit Statement for year 2 and 3 (associated with "inflated" Investment Base)

Year	NPV (Year 1)	1	2	3
(1) Realized net operational revenues			R$57.62	**R$81.82**
(2) Adjusted depreciation			R$50.00	**R$70.00**
(3) Interest (wacc) on non-depreciated capital (*)			R$10.00	**R$9.20**
Net Profit (= 1–2–3)	R$0.00	R$0.00	–R$2.38	R$2.62
	(0 = 0–2.38/1.1+2.62/1.21)			

(*) Calculated as the interest over the adjusted book value in the third year, added with the compounded interest charged over the increase in book value in the second year. That is: 10% X (R$ 120–R$ 50) + 10% X R$20 X 1,10

regarding investments and depreciation in the past could trigger escalation in the numbers used to guide capitalization and pricing decisions in the future.

5.3.4 Debt, water pricing and strategy: If you can't beat them join them?

In 2016 SEMASA contracted a consultancy study that was to provide contributions to the design of its business strategy (SEMASA, 2018). It estimated the economic value of the organization under three scenarios, each based on the projected discounted cash flow of net receipts (in present value of December 2017). In a "business as usual" base-scenario the company would continue as a municipal utility and maintain its existing portfolio of activities, resulting in a net worth of R$ 1,5 billion. The second scenario was similar but would devolve loss-generating operations to the municipality (solid waste; drainage and environmental management) and downsize to core business (water and sewage). This scenario would generate a net company worth of R$ 2,7 billion. A final scenario, generating economic value of around R$ 4,2 billion, would transform SEMASA into a mixed-capital company listed on the stock exchange, while maintaining municipal majority ownership. This increase in its economic value would emerge from a reduced cost of capital associated with an initial public offering of stock, considering it would avoid using expensive short-term working capital.[15]

Irrespective of the strategy it will adopt (at the time of writing still undecided) SEMASA is pressured by the courts and SABESP to settle its debt, which was estimated between R$ 2.7 and R$ 3.4 billion in December 2017. Considering the previous calculations, this practically rules out a scenario of "business as usual" (generating economic value of R$ 1.5 billion). Moreover, SABESP is keen on signing a concessionary contract with SEMASA, considering this brings the latter into the price-cap regime, provides more regulatory control and leverage of contracts via ARSESP and a more predictable revenue pattern for SABESP in the city.

Considering that the courts, ARSESP and anti-trust agency CADE point in the same direction (i.e. that debt should be paid in one way or another), this will either mean awarding SABESP a concessionary agreement against debt relief or restructuring debt and transforming SEMASA into a mixed-capital company listed on the capital market. Either way, this will require transformation in its strategy, including fragmentation in service delivery by devolving loss-generating activities to local government and increasing water prices.

In case of a concessionary agreement with SABESP, this will not include activities that generate financial burdens (environmental management, solid waste and storm drainage), also considering that the latter are not part of SABESP's core business.

Increasing water prices generates solutions (increasing financial viability and debt reduction) as well as problems, considering that SEMASA's empirical estimates for metropolitan São Paulo have shown that, under the present tariff structure, water pricing is notoriously regressive (SEMASA, 2018). For instance, in metropolitan São Paulo, income bracket 1 (monthly earning until R$ 830) and 7 (earning more than R$ 10,375) spend widely diverging percentages of income on water bills (4.58% versus 0.66% respectively). In the city of São Paulo, these discrepancies increase (6.39% versus 0.43%, respectively). To aggravate, price increases aimed at revenue maximization are likely to be more effective with low-income households, considering their reduced price elasticities of demand (reflecting their difficulty to substitute with alternative sources such as bottled water).

As discussed, SABESP's concessionary contracts with municipalities provide ample illustration of "stand and delivery" clauses aimed at maintaining its financial viability and reducing risks (Hildyard, 2016: 32–40; GESP and PMSP, 2010). Specific national sanitation data (Ministério das Cidades, 2018) indicate that concessionaire SABESP doesn't provide superior service delivery. In cities such as São Bernardo, which signed a contract in 2004, the share of families that faced service interruption has been consistently higher than in Santo André; during the water crisis in 2015, 98% of families in São Bernardo reported interruptions as compared to 38% in Santo André. The percentage of families that complained about service delivery in the city of Osasco, which signed its concession in 1999, has consistently surpassed the figure in Santo André (for example, in 2015, 17% versus 10% of households have registered complaints, respectively).

5.4 Conclusion

In addition to revealing a scenario whereby financialization of water in metropolitan São Paulo is still open-ended, our analysis provides two insights for the literature on urban financialization with relevance for the Global South.

First, in the absence of consolidated capital markets, fleshing out the entanglements between intergovernmental debts, pricing and valuation of infrastructure assets and water governance is key in understanding the making of financialization. Investigating how specific financial and institutional devices and pricing practices were set in motion sheds light on the gradual hollowing out of shared governance among state and local governments, and its filling in with elements of shareholder governance. We have briefly discussed the ongoing institutional-financial dispute between SEMASA and SABESP –increasingly driven by shareholder value premises – which has generated pressures to negotiate debts and a concessionary contract with the state company, implying increased water prices and devolution of less-profitable activity areas to general-purpose city government.

Santo André is not an isolated case in the Brazilian political economy of sanitation (Swyngedouw, 2013); in metropolitan São Paulo similar debt-triggered awards of municipal concessions to SABESP have led to an increasing penetration of shareholder governance in the cities of Osasco, São Bernardo, Diadema and São Paulo itself. While Mauá is still negotiating, in October 2018 the city of Guarulhos signed a preliminary concession agreement with SABESP. Moreover, changes in national legislation that stimulate private sector entrance in local government concessions are being discussed. Instead of the existing framework of Law No 11.445/2007, which allows for bilateral negotiation of concessions between state companies and local governments without tendering, the new design favours local governments to organize competitive bidding for concessions involving the private sector. Finally, the newly elected state government of São Paulo has announced plans to privatize SABESP, triggering a record increase in share prices to R$ 41.60 in January 2019.[16]

Second, political economy inspired analyses of what finance tends to do – transforming cities into bundles of tradable income-yielding assets and extracting value for that – can receive valuable inputs from social studies-oriented work aimed at investigating how financial markets are socially constituted in the first place through variegated entanglements between calculative practices and circulation of norms and conventions within professional communities, particularly considering the technological indivisibilities, risks, illiquidity and contestation that are inherent to city space (Pryke and Allen, 2019). We provided a first illustration of how these conceptual bridges could be established for metropolitan water finance and governance in São Paulo. Moreover, this cross-fertilization between political economy and social studies of finance could provide potential value-added for several of the themes in the literature on urban financialization we discussed before, such as the financialization of land; urban entrepreneurialism and the financialization through large urban development projects; and the nexus between capital markets and cities under austerity. Such work could set the stage for a research agenda according to which financialization is not an a priori entrance point for urban studies, but whereby the variegated and contradictory (re)production and appropriation of city space and collective wealth under the increasing influence of finance capital is embedded in a geographically and temporally specific reading of how fictitious markets for money (and credit), land and labour are socially constituted in the first place (Kirkpatrick, 2016).

Notes

1 Section 5.2, 5.3 and 5.4 are reprinted from the journal, *Urban Studies*, first online, Jeroen Klink; Vanessa Lucena Empinotti and Marcelo Aversa, "On contested water governance and the making of urban financialization: Exploring the case of Metropolitan São Paulo, Brazil", pages 6–20 (2019), with permission from Sage.
2 For a reference on Greater São Paulo see Empinotti et al. (2018).
3 This challenge remains considering the presidential veto on the creation of a national fund for metropolitan areas when approving the Statute of the Metropolis.
4 This court ruling has generated uncertainty regarding the specific financial design and governance model that will be adopted. Consequently, congress even prepared a proposal for a constitutional amendment on metropolitan governance in 2014, which has not advanced.
5 Based on the files of process 08700.011091/2015–18 within CADE (CADE, 2018).
6 In August 2017, Law No 16.525/(September 15, 2017) created a public holding company that is capitalized through the transfer of SABESP's assets and, eventually, by other capital emissions and transfers.
7 Based on general price index FGV. Available at: www.ipeadata.gov.br/ExibeSerie. aspx?serid=38390 (accessed 1 February 2019).
8 R$ 3.66 = 1 USD (as per 02/02/2019).
9 Margin squeezing is the reduction of the difference between the price of retail and the cost of wholesale water.

10 During 2007/08, *Lyonnaise des Eaux* was investigated for retail water margin squeeze in the province Ile de France.
11 CADE (2018). File 0132482.
12 CADE (2018). Nota técnica No 13/2017/CGAA3/SGA1/CADE.
13 CADE (2018). Nota técnica No 13/2017/CGAA3/SGA1/CADE: 12.
14 SEMASA's had requested access to SABESP's cash flow and accrual accounts, which was denied by CADE.
15 The economic value of assets is calculated as the net present value of expected income flows, using the average cost of capital in the discounting procedure. With given revenues, a lower cost of capital increases SEMASA's economic value. The projections indicated that working capital costs would rise to 11% in 2022, while an IPO could lower SEMASA's net cost of capital to around 8%.
16 The information is based on: Rocha R and Maia C (2019) Ação da SABESP volta a subir e bate recorde histórico na B3. *Valor*, 15 January.

References

Aversa M (2016) *História institucional do saneamento e da metropolização da grande São Paulo: trajetórias perdidas, conflitos inevitáveis.* Dissertação de mestrado. UFABC, São Bernardo do Campo, SP.

Bahl RW and Linn JF (1992) *Urban public finance in developing countries.* Washington: Oxford University Press.

Barcellos de Souza M (2013) *Variedades de capitalismo e reescalonamento espacial do Estado no Brasil.* Tese de Doutoramento. Instituto de Economia, UNICAMP, Campinas.

Birch K (2017) Rethinking value in the bio-economy: Finance, assetization and the management of value. *Science, Technology and Human Values* 42(3): 460–490.

Britto AL and Rezende SC (2017) A política pública para os serviços urbanos de abastecimento de água e esgotamento sanitário no Brasil: financeirização, mercantilizacão e perspectivas de resistência. *Cadernos Metrópole* 19(39): 557–581.

CADE (Conselho administrative de defesa econômica) (2018) *Pesquisa processual.* Processo 08700.011091/2015–2018. Available at: https://sei.cade.gov.br/sei/modulos/pesquisa/md_pesq_processo_exibir.php?0c62g277GvPsZDAxAO1tMiVcL9FcFMR5UuJ6rLq PEJuTUu08mg6wxLt0JzWxCor9mNcMYP8UAjTVP9dxRfPBcYwF7VPJpaiOt PaxB8YNJLtZLVCxY5rODYPctxCZ9E16 (Accessed 17 June 2017).

Callon M (1998) Introduction: The embeddedness of economic markets in economics: In: Callon M (ed.) *The law of markets.* Oxford: Blackwell, pp. 1–57.

Chiapello E (2015) Financialisation of valuation. *Hum Studies* 38: 13–35.

Christophers B (2014) From Marx to market and back again: Performing the economy. *Antipode* 57(1): 12–20.

Empinotti VL; Budds J and Aversa M (2018) Governance and water security: The role of the water institutional framework in the 2013–2015 water crisis in São Paulo. *Geoforum.* Online First. Available at: https://doi.org/10.1016/j.geoforum.2018.09.022 (Accessed 15 October 2018).

Ferreira LD (2019) Financiamento das políticas públicas de saneamento ambiental no Estado de São Paulo. In: *XVIII Encontro da ANPUR*, Natal, Brazil, May 27–31. Natal: ANPUR (Forthcoming).

GESP and PMSP (2010) State government and municipality of São Paulo. *Contrato de prestação de serviços públicos de abastecimento de água e de esgotamento sanitário.* Available at: www.saneamento.sp.gov.br/comitegestor/documentos/Contrato%20e%20 Convênio/Contrato_de%20_Prestação.pdf (Accessed 1 September 2018).

Guironnet A and Halbert L (2015) *Urban development projects, financial markets, and investors: A research note.* Chairville: École des Ponts Paritech.

Hall S (2010) Geographies of money and finance I: Cultural economy, politics and place. *Progress in Human Geography* 35(2): 234–245.

Henrique SM (2017) *A precificação dos serviços de saneamento de água e esgoto e o objetivo social.* Masters Dissertation. UFABC, São Bernardo do Campo.

Hildyard N (2016) *Licensed larceny.* Manchester: Manchester University Press.

Kirkpatrick LO (2016) The new urban fiscal crisis: Finance, democracy, and municipal debt. *Politics and Society* 44(1): 45–80.

Lapavitsas C (2013) *Profiting without producing: How finance exploits all of us.* London and New York: Verso.

MacKenzie D (2006) *An engine, not a camera: How financial models shape markets.* Cambridge, MA: The MIT Press.

Ministério das Cidades (2018) SNIS (*Sistema Nacional de informações sobre saneamento*). Available at: http://app3.cidades.gov.br/serieHistorica/ (Accessed 2–4 September 2018).

Moody's Investors Service (2013) *SABESP; Perspectiva estável.* Available at: www.sabesp.com.br/sabesp/filesmng.nsf/15652B2BF431F72083257BAA00719176/$File/PR_Sabesp_Port.pdf (Accessed 5 May 2018).

Peck J and Whiteside H (2016) Financializing Detroit. *Economic Geography* 92(3): 235–268.

Pryke M and Allen J (2019) Financialising urban water infrastructure: Extracting local value, distributing value globally. *Urban Studies* 56(7): 1326–1346.

Rezende F (2010) Em busca de um novo modelo de financiamento metropolitan. In: Magalhães F (ed.) *Regiões Metropolitanas no Brasil.* Washington: BID, pp. 45–98.

SABESP (2017) *Relatório Annual.* São Paulo: SABESP.

SEMASA (2018) *Avaliação econômico-financeira (Valuation).* Relatório apresentado à superintendência pela FGV. Santo André: SEMASA.

Swyngedouw E (2013) Águas revoltas. A economia política dos serviços públicos essenciais. In: Heller E and Esteban Castro JE (eds.) *Política pública e gestão de serviços de saneamento.* Belo Horizonte and Rio de Janeiro: Editora UFMG & Editora Fio Cruz, pp. 76–97.

Van Loon J; Oosterlynck S and Aalbers MB (2018) Governing urban development in the low countries: From managerialism to entrepreneurialism and financialization. *European Urban and Regional Studies*: 1–19. Available at: https://doi.org/10.1177/0969776418798673

Weber R (2010) Selling city futures: The financialization of urban redevelopment policy. *Economic Geography* 86(3): 251–274.

Zwan vd N (2014) Making sense of financialization. *Socio-Economic Review* 12: 99–129.

6 Building within the limits?

A closer look at the "My House,
My Life" Program in the outskirts
of metropolitan São Paulo

6.1 Introduction

This chapter analyzes what we will label as the "progressive" developmental
housing finance experience during the successive Worker's Party Administrations in federal government. More particularly, we discuss the flagship program,
Minha Casa Minha Vida (My House, My Life, abbreviated as MHML), which
was launched in 2009 in the midst of the international subprime crisis. The program represented an innovation considering that, for the first time in its housing
finance trajectory, there was a built-in progressive-redistributive grant targeted at
the poorest of the poor (i.e. those households earning up to three minimum salaries, which represented the so-called Target 1 component of the program).[1] At the
same time, however, the program revisited the previous experience of Brazilian
housing policy with credit-driven, private provision of social housing.

Although quite effective in terms of its short-run anti-cyclical effects on
income and employment generation, the program nevertheless became the object
of increasingly critical evaluations, particularly undertaken by academia.[2] Critics argued that the program threatened to repeat the dramatic experience of the
National Housing Bank (NHB) during the military regime (Cardoso, 2013; Rolnik et al., 2015; Ferreira, 2012), which was discussed in Chapter 3. According to
them, MHML was disconnected from the broader targets on housing deficit reduction (including its regional distribution) that had been discussed and approved
in the context of the national low-income housing policy. As such, MHML was
bound to produce excessive housing in regions with a low housing deficit (and
vice versa). The critics also argued that the program was repeating the NHB's pattern of provision in terms of building large-scale, isolated housing estates without
sufficient infrastructure and architectural quality, which were located on cheap
available land in the outskirts. Finally, the evaluations pointed out that MHML
seemed to be hollowing out the pillars of urban reform itself (Klink and Denaldi, 2015; Amore et al., 2015), structured around participatory master planning
and the application of redistributive planning instruments aimed at increasing the
leverage of municipalities over speculative land and real estate markets. In other
words, the credit-driven approach of social housing built by the private sector
was effectively exposing the regulatory deficiency of municipalities, which were

unable to contain the escalating land and housing prices in cities and metropolitan areas that were triggered by the increased liquidity in the first place.

On the basis of a specific case study on the impact of MHML in a subset of seven cities located in the industrial heartland of Greater São Paulo (the so-called ABC region),[3] our argument in this chapter will be somewhat different from this first generation of national evaluation studies on MHML. More specifically, although there was a built-in bias in detriment of reducing the housing deficit (which was only partially corrected in the second phase of implementation that started in 2012), any evaluation of the program's impacts should avoid general-izations and recognize the considerable geographic and historical differentiation among urban and city-regional trajectories.

In that sense, the particular experience of the ABC region with the program was emblematic. In the poorest target groups of the program (the so-called Target Group 1, with families earning up to three minimum incomes), which received significant budgetary grants in order to complement a small monthly installment, the program performed relatively better as compared to the outcomes from the national studies. This was partially due to the region's previous track record with the City Statute's redistributive planning tools such as inclusionary zoning, as well as the municipal investments that had been undertaken in slum upgrading in the periphery since the 1980s. Thus, in the specific experience of the region, Target Group 1 projects had better locations and were also accompanied by a relatively dense network of complementary infrastructure in its more peripheral regions.

At the same time, however, the program was facing challenges in the region in relation to its scheme for more affluent target segments (i.e. Target Groups 2 and 3, earning between four to ten minimum salaries), which operated with subsidized credits instead of outright grants and required less support from local government. As a matter of fact, as will be discussed subsequently, in this scheme the developers simply removed part of the units that had initially been planned and contracted according to the prevailing finance guidelines of the program, with pre-determined price ceilings and an amount of subsidies for Target Groups 2 and 3, and sold these to mid-income groups in order to reap the financial gains associated with escalating land and real estate prices that accompanied the boom-ing Brazilian economy. In other words, the built-in flexibility of the program – allowing the sale of units, at the end of construction, "outside the established limits" regarding the income level of target groups and price levels of housing units – as well as the lack of progress in effectively implementing the City Statute by combatting speculative urban markets, enabled developers and land owners to capitalize subsidies into their pricing and valuation practices. This ended up reducing the effectiveness of the program itself in its more affluent Target Group 2 and 3 segments. In a way, then, it was not only the project-driven approach of MHML that threatened to bypass the complex, longer-term dynamics of urban reform; at the same time, the lack of structural leverage of local governments over speculative land markets also led to a gradual hollowing out of the program's effective capacity to target and deliver credit-driven social housing itself, more particularly in a setting where it was supposed to be built by the private sector for specific socioeconomic segments.

After this introduction, the chapter is organized in three complementary sections. In the first we provide a brief description of the program's anti-cyclical character and its overall design (subsidies and price ceilings structured around three target groups (1 versus 2 and 3); the role of private developers in the higher-income segments (Target Groups 2 and 3, respectively) and local governments in the lower-income tranche (Target Group 1). Moreover, we situate the emergence of the program in the context of the earlier-discussed restructuring and rescaling of the Brazilian developmental state and the role of financialization. Finally, in order to prepare the reader for the case study, we review the main critique received by the program.

In a subsequent section we evaluate the program's impact on the ABC region. After a brief characterization of the region and its real estate market, we investigate the program's impact in the region and discuss the similarities and differences between our findings and the national evaluation studies.

On the basis of the empirical work, the concluding section then argues that the rise (as well as the decline) of MHML provides important elements for a research agenda on the design of progressive-redistributive urban and housing policy itself, particularly in a scenario marked by an increasing penetration of financial logics and calculative practices into the pricing, subsidization and valuation of land and real estate in contemporary Brazilian cities (DiPasquale and Wheaton 1996).

6.2 My House, My Life – Initial design, limits and critical evaluations

6.2.1 A primer on the program

In a way, it can be argued that the launch of the ambitious MHML program symbolized the hybridity of the social-developmental project itself during the successive federal administrations of the Worker's Party during 2003–2015. Implemented as a rapid Brazilian response to the impact of the international subprime crisis on its economy, it was not only designed as a counter-cyclical shock aimed at income and employment generation through the building and construction sector but also as an innovative approach to the housing deficit through the large-scale provision of affordable housing by the market through outright budgetary grants as well as subsidized credit (Klink and Denaldi, 2014).

Arcanjo (2016) is correct in claiming that MHML in reality represents two programs that have been built into one. Its Target Group 1 component represents the real innovation. It emerged as part of a historical recognition that the poorest households (earning up to three minimum salaries, concentrating around 91% of the national housing deficit) couldn't afford market-based housing and, as such, required direct support from the public fund, as well as a more active role for local government in the design and implementation of projects. The second component of the program, focussed on Target Group 2 (families earning between four and six minimum salaries per month) and 3 (earning between seven and ten minimum salaries) was more traditional, as reflected in the utilization of the compulsory savings fund from workers from the HFS – the FGTS – in order to stimulate the private provision of social housing through subsidized suppliers' and buyers' credit, backed up by a guarantee fund.

In this chapter we focus on Phase I and II of MHML, before the drying up of the liquidity that has affected the program as a result of the ongoing economic and political crises since 2015.[4] Up to the year 2012, 1,005,128 units were contracted. Having reached its initial target of 1 million houses, a second phase was launched in September 2012, and projected another 2.4 million units, of which 1.6 would be allocated to Target Group 1 (Magalhães, 2013; Brasil, 2013).

The institutional and financial design of MHML cannot be dissociated from the broader international circulation of ideas and concepts regarding the need to simultaneously reduce direct intervention in housing markets, create financially viable mechanisms for credit and, finally, target more effectively the available subsidies to the poorest households (Freitas et al., 2015; Hoek-Smit and Diamond, 2003; World Bank, 1993). As such, MHML provides targeted and income-related support through lump-sum grants, subsidies and cheap credit to the demand and supply side of the market, all within pre-established limits regarding the prices for finished housing units (see Tables 6.1 and 6.2 for the specific ceilings regarding house prices and the socio-economic profile of potential buyers for Target 1 and 2/3, respectively). Although local government is more active in the design of projects in the Target 1 range of the program, MHML clearly delegates the development and building of units to the private sector. In order to succeed with this strategy, reduction of systemic risk for developers and builders is a key element. In that respect, MHML delivers "value for money", considering the significant budgetary grants and the involvement of local governments in land and infrastructure delivery with Target Group 1 projects, while partial subsidization (lump-sum capital grants and reduced interest payments), a public guarantee fund that covers loan defaults during the construction period, as well as transfer of the ownership of units to the national housing bank at the end of construction practically eliminate risks under the Target 2 and 3 tranches.

The Target Group 1 scheme of MHML was actually implemented by the Lula administration by giving a creative, more redistributive twist to a fund that was originally designed in the 1990s as a market-enabling device aimed at stimulating rent-to-buy options linked to credit. To be specific, the so-called Residential Leasing Fund was effectively filled in according to a more progressive-redistributive project, which enabled target group members to become owner-occupiers after having paid heavily subsidized, non-interest-bearing monthly instalments during 10 years, under the condition of having properly maintained their units. In that scenario, effective subsidies varied from 60% to almost 90% of the final price of units (Brasil, 2013), or the present value equivalent of the series of required monthly payments of 5% of available income, or R$ 50.00, during ten years.

Table 6.1 MHML – Target group 1 – Price limits per region (phase I)

State	Territories	Price Limits for Aquisition of Units	
		Apartment	House
Federal District of Brasilia	Capital	76,000.00	76,000.00
	Municipalities of the Integrated Development Network of the Federal District and its surroundings (RIDE region)	60,000.00	60,000.00
Goiás, Mato Grosso Do Sul and Mato Grosso	Capital city and Metros	60,000.00	60,000.00
	Municipalities with more than 50.000 inhabitants	57,000.00	57,000.00
	Municipalities with less than 50.000 inhabitants	N/A	56,000.00
Bahia	Capital city and Metros	64,000.00	64,000.00
	Municipalities with more than 50.000 inhabitants	60,000.00	60,000.00
	Municipalities with less than 50.000 inhabitants	N/A	57,000.00
Ceará and Pernambuco	Capital city and Metros	63,000.00	63,000.00
	Municipalities with more than 50.000 inhabitants	59,000.00	59,000.00
	Municipalities with less than 50.000 inhabitants	N/A	56,000.00
Alagóas, Maranhão, Paraíba, Rio Grande do Norte and Sergipe	Capital city and Metros	61,000.00	61,000.00
	Municipalities with more than 50.000 inhabitants	57,000.00	57,000.00
	Municipalities with less than 50.000 inhabitants	N/A	54,000.00
Piauí	Capital city	61,000.00	61,000.00
	Municipalities with more than 50.000 inhabitants	57,000.00	57,000.00
	Municipalities with less than 50.000 inhabitants	N/A	54,000.00
Acre, Amazonas, Amapá, Pará, Rondónia, Roraima and Tocantins	Capital city and Metros	62,000.00	62,000.00
	Municipalities with more than 50.000 inhabitants	60,000.00	60,000.00
	Municipalities with less than 50.000 inhabitants	N/A	58,000.00
Espírito Santo	Capital city and Metros	60,000.00	60,000.00
	Municipalities with more than 50.000 inhabitants	58,000.00	58,000.00
	Municipalities with less than 50.000 inhabitants	N/A	56,000.00
Minas Gerais	Capital city and Metros	65,000.00	65,000.00
	Municipalities with more than 50.000 inhabitants	60,000.00	60,000.00
	Municipalities with less than 50.000 inhabitants	N/A	58,000.00

Rio de Janeiro	Capital city and Metros	75,000.00	75,000.00
	Municipalities with more than 50.000 inhabitants	69,000.00	69,000.00
	Municipalities with less than 50.000 inhabitants	N/A	60,000.00
São Paulo	Municipalities part of the metropolitan regions of the capital city, Campinas, Santos and Jundiaí.	76,000.00	76,000.00
	Municipalities with more than 50.000 inhabitants	70,000.00	70,000.00
	Municipalities with less than 50.000 inhabitants	N/A	60,000.00
Rio Grande do Sul, Paraná and Santa Catarina	Capital city and Metros	64,000.00	64,000.00
	Municipalities with more than 50.000 inhabitants	60,000.00	60,000.00
	Municipalities with less than 50.000 inhabitants	N/A	59,000.00

Source: Author's elaboration. Based on Internal Decree 168, April 12th, 2013. Ministry of Cities.

Table 6.2 Limits and price ceilings – MHML target group 2 and 3 projects

Gross Family Income	Target Group	Max. % Installment/income	Interest rate	Price Ceiling Housing Unit	Loan to value ratio
Until R$ 2,455	1	30%	5.00%	R$190,000	90%
From R$ 2,455 to R$ 3,275	2	30%	6.00%	R$190,000	90%
From R$ 3,275 to R$ 5,000	3	30%	7.16%	R$190,000	90%

Source: Author's elaboration. Based on Arcanjo (2016: 48).

Local governments are supposed to participate actively in Target 1 projects, both through the registering and organizing of beneficiary cadastres and the mobilization of land and complementary infrastructure in order to generate financially viable projects within the limits established for price and subsidy ceilings. In that respect, cities with a track record of actively using and filling in their land-use planning according to redistributive-progressive premises, for example, through the provision of affordable and well-located zones for low-income housing in the master plans (for instance, through so-called Special Social Interest Zones, as envisaged by the federal City Statute), have shown a better performance in terms of finding good locations for Target 1 projects.

Target Group 2 and 3 projects have less involvement of local governments, in general no direct federal budgetary grant money and operate through subsidized credit. Yearly interest rates on loans to families sourced by the FGTS fund rise proportionally with family income (from 5% to 7.16%). Nevertheless, depending on income and geographical location, a direct subsidy of up to R$ 25,000 is available for Target 2 projects. During construction period, final beneficiaries sign loan contracts for up to 30 years, while building companies, if they chose to do so, can also finance up to 85% of developments.

6.2.2 *Emerging critical evaluations*

At the completion of its first phase, a series of critical evaluations emerged regarding the impact of the program on Brazilian urban and rural territories (Cardoso, 2013; Ferreira, 2012; Krause et al., 2013; Marques and Rodrigues, 2013). Part of this critique should be seen in light of the fact that the program was launched when the National Housing Plan, with specific proposals for the finance and implementation of social housing, was still in its final stage of elaboration. The critics argued that the launch of MHML had effectively hollowed out the energy and attention spent on the elaboration of the plan.

Before fleshing out the more specific arguments that came out of the evaluations, however, it should be recognized that MHML actually did follow the Housing Plan's recommendation for targeted and budgetary resources for the lowest income groups. As mentioned, by doing so, it explicitly innovated in relation to Brazil's historical trajectory of housing finance by recognizing that Target Group

1 couldn't afford market-based housing, and, as such, would need significant back-up from the public fund.

The evaluation studies raised three critical contradictions in the program.

First, although the program introduced budgetary grants and significantly raised the amount of subsidies, MHML didn't reach the traditionally excluded low-income groups of Target 1 projects. As a matter of fact, the program's initial design reflected a mismatch considering that it only allocated 40% of its resources to Target 1 projects, whereas 91% of the national housing deficit was concentrated in this segment. The second phase only partially corrected this distortion by increasing overall allocation to 60%. As a consequence, the program actually overshoots the already-oversized targets for low-to-medium income groups 2 and 3, while it underperforms in relation to Target 1 projects (See Table 6.3).[5] Second, the design of MHML regarding the allocation of the available resources to cities and regions also didn't follow the guidelines of the National Housing Plan, which was based on geographical criteria in terms of the distribution of the housing deficit within Brazil. As a result, cities with high deficits received too few units, while in smaller municipalities housing production effectively exceeded the deficit, generating additional population growth and operation and maintenance expenditures that burdened small local budgets.

Third, the fast-track, project-driven approach of the program had effectively hollowed out the more complex agenda that was structured around the participatory-collaborative planning aimed at the elaboration of progressive-redistributive local master plans, which were also the backbone of the National Housing Plan's architecture. As such, the critics argued that the program bypassed the political project of urban reform that was aimed at increasing the leverage of local governments over speculative land and real estate markets through the effective implementation of the National City Statute.

As a matter of fact, the bulk of the evaluation studies prioritized Target Group 1 projects and pointed out that the escalating land prices of overheated real estate markets, given the limits for subsidies and price ceilings, effectively generated a pattern of peripheral locations with deficient infrastructure, particularly in smaller cities outside metropolitan regions. The studies also indicated similarities with the NHB trajectory, in the sense that the additional credit, as well as the gains in production and productivity, didn't result in better and more affordable housing. Instead, there was evidence that the grants, subsidies and tax incentives were being capitalized by developers and construction companies, effectively reinforcing successive cycles of land price escalation, higher profit margins and lower quality of increasingly standardized units (Ferreira, 2012; Mendonça and Sachsida, 2012; Denaldi, 2013). Not unsurprisingly, the limits for subsidy guidelines and price ceilings within the program were adjusted regularly. A study undertaken by Eloy et al. (2013: 16), for example, showed that between 2009–2012, only within the Target 1 projects, the real value of subsidies had increased by 28%.

In short, while it was recognized that MHML innovated in terms of recognizing the need for stable and predictable support from the public fund in order to

Table 6.3 Accumulated housing deficit X national targets and performance of MHML: Phase I, II and III

	Brazil: Accumulated Housing Deficit (%)	Brazil: Accumulated Housing (Units)				
Target 1 projects	91%	6,550,000				
Target 2 projects	6%	430,000				
Target 3 Projects	3%	210,000				
	100%	7,190,000				
	MHML Phase I (*)		*Contracted*			
	Target (Units)	%	R$ (Billions)	%	Units	%
Target 1 projects	400,000	40%	18	33%	482,741	48%
Target 2 projects	400,000	40%	26.1	47%	375,764	37%
Target 3 Projects	200,000	20%	11	20%	146,623	15%
	1,000,000	100%	55.1	100%	1,005,128	100%
	MHML Phase II (**)		*Contracted*			
	Target (Units)	%	R$ (Billions)	%	Units	%
Target 1 projects	1,600,000	67%	63.2	34%	1,226,605	45%
Target 2 projects	600,000	25%	100.4	53%	1,216,341	44%
Target 3 Projects	200,000	8%	24.7	13%	307,054	11%
	2,400,000	100%	188.3	100%	2,750,000	100%
	MHML Phase III (***)		*Contracted*			
	Target (Units)	%	R$ (Billions)	%	Units	%
Target 1 projects	170,000	30%	2	2%	53,748	2%
Target 2 projects	200,000	35%	86.5	82%	715,742	26%
Target 3 Projects	200,000	35%	17	16%	128,110	5%
	570,000	100%	105.5	100%	897,600	100%

Source: Author's elaboration. Based on Pereira (2019: 152).

(*) April, 2009. Family Income up to R$ 4,650
(**) September, 2012. Family Income up to R$ 5,000
(***) March, 2016. Family Income up to R$ 6,500

provide housing for the poorest households, it had missed an important opportunity to link up with the project of urban reform and intervention in land markets. In the next section we will analyze a detailed impact study for the ABC region and discuss its results in light of these national evaluations.

6.3 My House, My Life in the outskirts of metropolitan São Paulo: An exploration of the Greater ABC region[6]

This case study section is organized in two complementary parts. The first provides an overview of the socio-economic profile of the ABC region and its real estate dynamics in the first decade of the 2000s. We discuss the gradual macroeconomic recovery and its relations with the real estate financial dynamics in the city of São Paulo and its metropolitan hinterland, with a particular zoom on its unfolding in the ABC region. In the second part we will flesh out the impacts of MHML on the region, with a priority on the investigation of its entanglements with the local real estate markets and the location-insertion of projects (availability of infrastructure, community equipment and so on) into the urban fabric, particularly in light of the earlier-mentioned critique from the national evaluation studies.

6.3.1 A characterization of the ABC region and its real estate market in the first decade of the 21st century

The ABC region is composed of seven cities with a total population of a bit less than 3 million inhabitants. It is located in the southeast of the metropolitan region of São Paulo (see Figure 6.1). The demographic profile of the region is summarized in Table 6.4.

To some extent, the region is a symbol of Brazil's national development regime; its first industrial take-off occurred in the early-twentieth century when the country was making its transition from an export-led regime to a national industrial space economy through the gradual buildup of a sector for non-durable consumer goods (such as textiles, food and beverage). In a subsequent stage, the metropolitan region of São Paulo and ABC concentrated the bulk of the national car manufacturing and petrochemical industry that was being set up during the period 1930–1970 (see also Chapter 3). The demise of the national developmental regime and the productive restructuring in industrial sectors such as car manufacturing dramatically affected the region in the 1990s (Rodríguez-Pose et al., 2001). Table 6.5, based on the numbers from the Ministry for Work and Employment, illustrates the effect of the industrial downsizing that occurred since then on employment levels in Brazil, the state of São Paulo and the metropolitan capital region, including the ABC region. Although the macroeconomic growth during 2003–2013 helps to recover some of these employment losses, the figures never return to the levels that were prevailing at the end of the 1980s. After 2013, industrial employment in Brazil and metropolitan São Paulo has been affected dramatically. Theoretically, part of these employment losses could be attributed to growth in industrial productivity and efficiency. Table 6.6 nevertheless shows that the value added generated by the region (a proxy for income generation) during 2012–2016 has shown a yearly average reduction of 4,07%. Particularly the cities of São Bernardo do Campo and São Caetano do Sul, which concentrate the bulk of multinational car manufacturing in the region, have shown impressive yearly

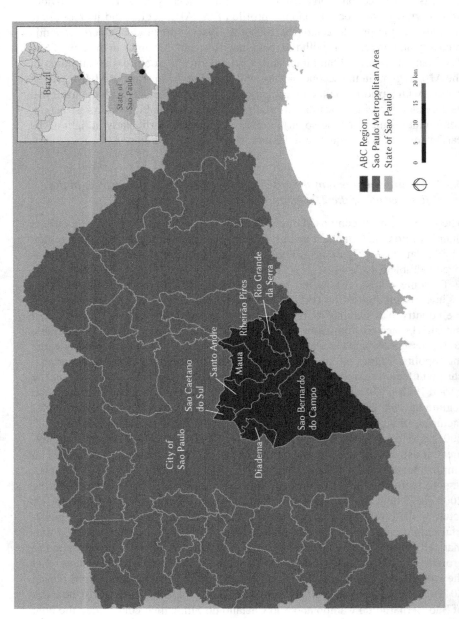

Figure 6.1 The location of the ABC region in metropolitan São Paulo, the state of São Paulo and Brazil

Table 6.4 Population of metropolitan São Paulo, São Paulo city and the ABC region (2019)

Metropolitan São Paulo	20,996,747
Municipality of São Paulo	11,811,516
ABC Region:	2,677,743
Diadema	402,813
Mauá	456,020
Ribeirão Pires	118,441
Rio Grande da Serra	49,229
Santo André	692,207
São Bernardo do Campo	807,917
São Caetano do Sul	151,116

Source: Author's elaboration. Based on Fundação SEADE. Informações dos Municípios Paulistas. Available at: www.imp.seade.gov.br/frontend/#/tabelas (accessed on August 15, 2019).

Table 6.5 Evolution of formal industrial employment per region (1989–2017)

	Industrial Employment					
	1989	*1992*	*1999*	*2006*	*2013*	*2017*
São Paulo (Mun.)	1,024,350	705,672	475,879	526,608	539,543	404,329
ABC Region	363,333	256,183	187,759	232,558	258,550	182,168
Metropolitan São Paulo	1,751,740	1,238,122	902,500	1,087,809	1,174,154	898,251
State of São Paulo	2,852,008	2,115,893	1,756,312	2,369,974	2,822,589	2,330,177
Brazil	6,151,654	4,713,262	4,603,893	6,594,783	8,292,739	7,105,206

	As a percentage of total Brazilian formal Industrial Employment					
	1989	*1992*	*1999*	*2006*	*2013*	*2017*
São Paulo (Mun.)	16.65%	14.97%	10.34%	7.99%	6.51%	5.69%
ABC Region	5.91%	5.44%	4.08%	3.53%	3.12%	2.56%
Metropolitan São Paulo	28.48%	26.27%	19.60%	16.49%	14.16%	12.64%
State of São Paulo	46.36%	44.89%	38.15%	35.94%	34.04%	32.80%
Brazil	100.00%	100.00%	100.00%	100.00%	100.00%	100.00%

Source: Author's elaboration. Based on Ministry of Work and Employment (MTE). Rais-CAGED.

reductions in value added of 6,33% and 7,16%, respectively. Although not in the same dramatic manner, metropolitan São Paulo is also losing value added during the same period. At the same time, since the 1990s Brazilian real estate markets have gone through a series of transformations. Part of these were influenced by state regulatory restructuring regarding housing and the built environment (Royer, 2014; Shimbo, 2012), such as the earlier-mentioned creation of the Real Estate Finance System (1997), the strengthening of fiduciary alienation, the flexibilization of the restrictions regarding the use of FGTS funds for real estate investments and the release of guidelines for the floating of building and construction companies and developers on the stock exchange, among some of the more important measures (See Chapter 3).

Table 6.6 Value added of metropolitan São Paulo, São Paulo city and the ABC region (2012–2016) (in real prices of 2016)

	2012	2013	2014	2015	2016	Average annual growth (2012–2016)
Metropolitan São Paulo	R$923,600,595	R$936,361,710	R$960,233,559	R$960,560,419	R$914,410,060	−0.25%
Municipality of São Paulo	R$573,249,464	R$576,494,931	R$586,800,437	R$597,469,016	R$569,910,503	−0.15%
ABC Region:	R$111,601,448	R$113,379,355	R$109,610,009	R$99,702,963	R$94,506,240	−4.07%
Diadema	R$12,773,947	R$13,036,008	R$13,283,933	R$12,441,526	R$10,997,442	−3.67%
Mauá	R$9,936,828	R$10,468,015	R$10,792,401	R$11,173,566	R$12,017,266	4.87%
Ribeirão Pires	R$2,558,029	R$2,757,609	R$2,797,611	R$2,831,652	R$2,706,251	1.42%
Rio Grande da Serra	R$533,673	R$585,322	R$545,925	R$540,098	R$536,006	0.11%
Santo André	R$25,890,004	R$27,011,897	R$26,282,628	R$24,080,973	R$22,523,809	−3.42%
São Bernardo do Campo	R$45,509,060	R$45,301,867	R$42,787,068	R$37,021,938	R$35,028,835	−6.33%
São Caetano do Sul	R$14,399,907	R$14,218,636	R$13,120,443	R$11,613,210	R$10,696,632	−7.16%

Source: Author's elaboration. Based on Fundação SEADE. Informações dos Municípios Paulistas. Available at: www.imp.seade.gov.br/frontend/#/tabelas (accessed on August 15, 2019).

Moreover, any evaluation of MHML should take into consideration the macroeconomic recovery that was occurring in the country from 2003 onwards, as well as the impact of state regulatory restructuring on the real estate dynamics in metropolitan São Paulo in general, and in the ABC region in particular. Therefore, before "zooming in" on the specific impacts of the MHML program itself, the rest of this section will present an outline on what was occurring in the overall real estate sector of the ABC region during 2000–2011, on the basis of a quantitative and qualitative analysis of new development that was undertaken by Klink et al. (2016) and UFABC (Universidade Federal do ABC) (2016).[7]

Figure 6.2 shows the number of new housing units developed during the period analyzed (2000–2011), according to the source of housing finance. More specifically, developments were financed by the social housing fund FGTS or *Caixa Economica Federal* (being the equivalent of the National Housing Bank and mainly working with FGTS funding), the Brazilian savings and loan system (SBPE) – targeted at medium income and market-based housing – private banks or cases that didn't provide any information. The latter category was composed of funding from non-governmental sources, such as reinvested internal funds from developers and construction companies or up-front payments from buyers. This category showed a systematic fall along the period analyzed; while during 2003–2006 non-governmental sources of finance (reinvested funds and/or upfront payments of instalments from buyers) represented more than half of new development, the market share of this source fell to less than one-third in 2011. In other words, the macroeconomic recovery, which automatically generated an increase in FGTS funding (considering the latter is based on a surcharge on wages), as well as the launch of MHML, triggered an increase in state-driven sources of credit (and a proportional reduction of private credit).

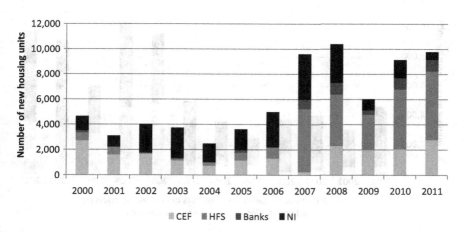

Figure 6.2 Number of new housing units in the ABC region according to source of funding (2000/11)

Source: Author's elaboration. Based on Klink et al. (2016).

An important trend break occurred in 2006–2007, mainly in light of the increased activity of the HFS system (FGTS and CEF). The sector expanded from 846 new units in 2006 to 4,988 new dwellings in 2007. The volume of private bank credit was also growing, as can be seen in the increase from 60 to 741 new units during the same period. Moreover, the overall market in the ABC region underwent a significant increase in scale, considering that the number of new units developed grew from 4,992 in 2006 to 9,584 in the year after. As mentioned, during this period a new legal framework was enacted, which facilitated the floating of stock of developers and building-construction companies in the capital market.

In Figure 6.3 the same figures on new development are presented according to the capital structure of firms, that is, whether developers are public, listed companies or privately held entities. The pattern that emerged reinforces the previous figure. Until 2007, the market share in new development of public companies was consistently below 25%, with the exception of the year 2004, which was marked by relatively low business activity. After 2007, this share rose to more than 50% and never fell below 33%.

Another distinguishing feature is that, at first sight, public companies were better prepared to respond to the incentives provided by MHML and the renewed macroeconomic growth in 2010: while these firms increased new developments from 3,240 in 2009 to 5,965 units in 2010, privately held companies showed a smaller increase in the same year (i.e. from 2,780 to 3,524 new units).

Figure 6.4 shows new development according to the geographical location of the headquarters of developers, that is, whether the main office was located in the ABC region, the city of São Paulo, in other cities or whether new developments

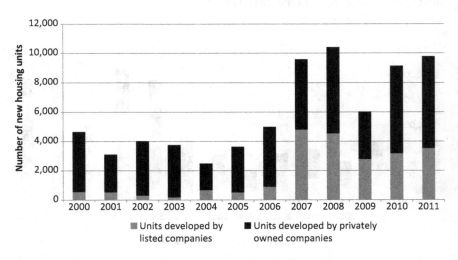

Figure 6.3 Number of new housing units in the ABC region by listed and privately owned companies (2000/11)

Source: Author's elaboration. Based on Klink et al. (2016).

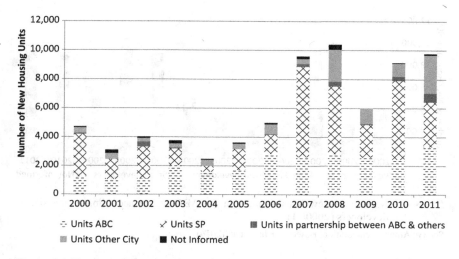

Figure 6.4 Number of new housing units in the ABC region according to location of headquarters of developers (2000/11)

Source: Author's elaboration. Based on Klink et al. (2016).

emerged in the context of a partnership between local firms (from the ABC region) and external companies (outside the ABC region).

In 2007, the market share of developers with head offices based in the city of São Paulo in total real estate activity in the ABC region increased to almost two-thirds, as compared to their more modest figures in previous years. This reinforces the hypothesis, concerning the real estate market in ABC, of a correlation between the activity of listed, public companies (Figure 6.3) and the presence of developers with headquarters in the city of São Paulo, considering the latter's similar growth pattern from 1,322 new units in 2006 to 6,244 new dwellings in the next year.

Companies with headquarters in São Paulo showed more volatility in new development in comparison with local firms. For instance, the former launched a record of 6,244 new developments in 2007, while these numbers fell in the subsequent two years (that is, to 4,694 in 2008 and 2,247 in 2009). In 2010, driven by the effect of MHML and the macroeconomic growth figures, new development again grew to a significant figure of 5,466 units. The fluctuations were much smoother for local firms during 2006–2011, considering that the difference between the best and worst year was never beyond 734 units launched.

Figure 6.5 shows new development according to the size of companies (large, medium, small).[8] Until 2006, in each year the market share of big developers in the ABC region never exceeded one-third of new development. In 2007, the share of these companies in the total flow of new real estate increased to more than 50%, while it grew to approximately two-thirds of new development in the region during 2007–2009.

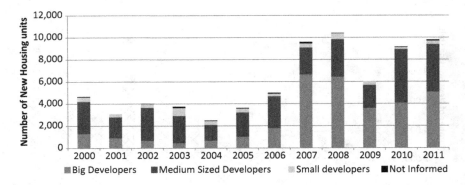

Figure 6.5 Number of new housing units in the ABC region according to the size of developers (2000/11)

Source: Author's elaboration. Based on Klink et al. (2016).

Finally, in Figure 6.6 we show the relation between the growth of new developments (according to the source of finance) and the evolution of real estate prices in the ABC region during the period 2000–2011. All prices per square meter were calculated in real terms of the base year 2011.[9,10] There are clearly two different periods. During 2000–2004, useable per-square-metre prices fell from R$ 3,000/m^2 to around R$ 2,250/m^2. After that year, prices consistently rose until reaching values of around R$ 4,250/m^2, and were also accompanied by an increase in scale (from 6,000 new units in 2004 to more than 11,000 units in 2011) and a qualitative restructuring of the market itself, as discussed when we analyzed the entrance of new players (bigger developers, frequently registered as public companies and some of them headquartered in the city of São Paulo).

6.3.2 An evaluation of the impact of MHML on the ABC region

6.3.2.1 General characterization of MHML in the region

In Figure 6.7, we illustrate the evolution of the numbers produced by the program in each of the cities during 2009–2014 (therefore including part of the production generated during the second phase of the program). During this period, the program managed to contract 19,222 units in the ABC region. The year 2012 generated the highest volume, considering the program contracted 5,075 units. All contracts were exclusively concentrated in the four bigger cities of the ABC region, that is, Santo André, São Bernardo do Campo, Diadema and Mauá, with the higher figures for São Bernardo do Campo with 5,837 units, while the cities of Mauá, Santo André e Diadema contracted 5,278, 4,998 and 3,174 units, respectively.

In Figure 6.8, the evolution of the number of units contracted in the same period is shown according to the main target groups of the program.

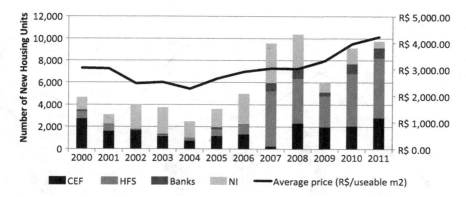

Figure 6.6 Number of new housing units in the ABC region according to source of funding and average price/useable m^2 (2000/11)

Source: Author's elaboration. Based on Klink et al. (2016).

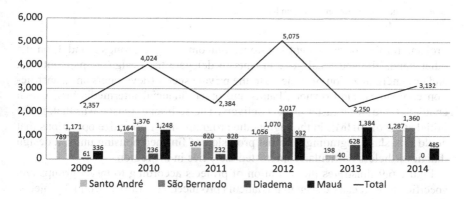

Figure 6.7 MHML: Evolution of the number of housing units contracted in the ABC region by city (2009–2014)

Source: Author's elaboration. Based on Klink et al. (2016).

In general, the number of units contracted for Target 1 projects in the initial years was small, but this tranche gradually increased its market share in the overall housing production, particularly from 2012 onwards. During the period analyzed, Target Group 1 projects represented a market share of a little over one-third (that is 37%) of the units contracted by the program in the region.

The slower evolution of Target Group projects was due to two reasons. First, this modality required a more active and time-consuming involvement of local governments in terms of articulating projects with overall housing and urban development policies, in addition to the organization of the cadastre of potential target groups and, if necessary, the mobilization of land and infrastructure.

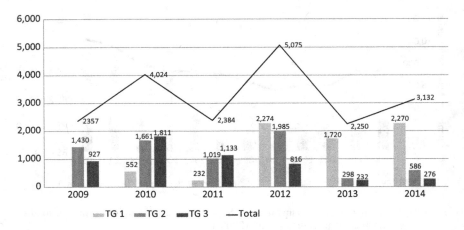

Figure 6.8 MHML: Evolution of the number of housing units contracted in the ABC
region by Target Groups (2009–2014)

Source: Author's elaboration. Based on Klink et al. (2016).

Projects for the more affluent low-to mid-income Target Groups 2 and 3, on the
other hand, allowed local governments to delegate the design, implementation
and commercialization of projects to the private sector (developers and construction companies, land owners, banks and other financial intermediaries), which
enabled a somewhat faster project delivery. The additional reason was that, in the
initial stage of MHML in the region, the program didn't prioritize operations that
could link slum upgrading with the provision of finished housing units, a design
that nevertheless particularly interested the cities in the region.[11]

Figure 6.9 illustrates the evolution of projects according to target groups and
specific cities. It can be seen that Diadema produced few Target Group 3 projects.
At the same time, in Santo André the market share of Target Group 1 units was
relatively higher than in the other cities of the region.

Figure 6.10 summarizes the earlier mentioned (mis)match between effective
housing units contracted and the general goals to be reached for reducing the
housing deficit in the region, according to specific Target Groups (1, 2 and 3).
In spite of the higher absolute figures for Target Group 1 projects, its relative
performance in terms of contributing to the reduction of the housing deficit
is less effective (7,048 units out of a deficit of 58,962, that is, a contribution
of 11.95%). In other words, the performance of the program in terms of the
reduction of deficits for Target 1 is relatively lower when compared with its
contribution to reducing the (smaller) deficit in the Target 2 and 3 communities
(that is, a production of 6,979 units for a housing deficit of 14,419 in the Target
Group 2 range and 5,195 units for a housing deficit of 15,950 for Target Group 3
beneficiaries).

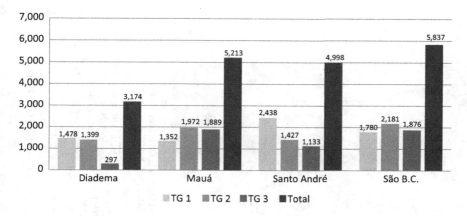

Figure 6.9 MHML: Evolution of the number of housing units contracted in the ABC region by Target Group and city (2009–2014)

Source: Author's elaboration. Based on Klink et al. (2016).

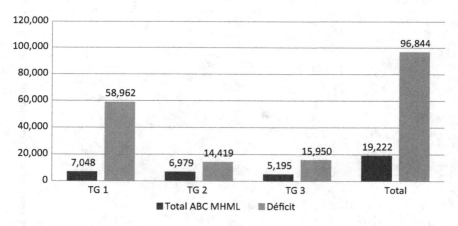

Figure 6.10 MHML: Contribution to reduction of the housing Deficit in the ABC region, by Target Group (2009–2014)

Source: Author's elaboration. Based on Klink et al. (2016).

6.3.2.2 Location and availability of infrastructure

With the purpose of comparing the location of MHML units within the overall real estate dynamics in the ABC region, which was analyzed in the previous section, Figure 6.11 shows the location and price ranges of new real estate developments for the specific year 2013 (according to the CEM-EMBRAESP data base) and the distribution and price range of all units of the MHML program, during 2009–2014.[12]

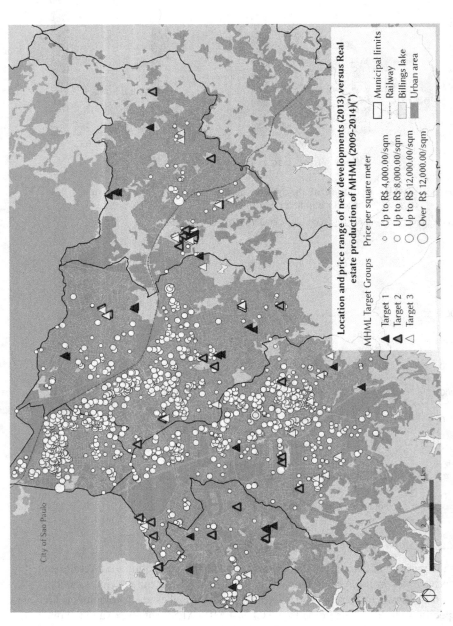

Figure 6.11 Location and price range of new developments (year 2013) versus real estate production of MHML (period 2009–2014)

Source: Author's elaboration. Based on CEF (2015).

This figure shows two patterns. The first is a relative concentration of new real estate development in centrally located areas of São Caetano do Sul, São Bernardo and Santo André, which have all gone through an escalation of real estate prices. Part of this new development occurs in well-located older industrial corridors, which are either going through a process of productive restructuring (leaner systems and so on) or outright downsizing or closure (recall the continuous decline in industrial employment, Table 6.5), and provide concrete perspectives for the additional supply of land in an increasingly overheated real estate market.[13] Moreover, in light of the prices charged in these older industrial corridors, it is unsurprising that, with the exception of a few Target 2 and 3 projects, most of the housing provision by MHML is somewhat distant from this more centrally located real estate dynamics. However, as will be discussed, distance is not necessarily equivalent to precariousness.

Although there is indeed a tendency that Target Group 1 projects are further away from city centres, nearby the highways, main access roads or slum settlements, there are, nevertheless, several projects that have been implemented in good locations. The reason is that some of these projects were approved and matched with existing Special Social Interest Zones designed for low-income families, *Zonas Especiais de Interese Social* (ZEIS), either on public or private property. Special Social Interest Zones (ZEIS) are part of the master plan that requires approval by the municipal council. If they have reasonable locations, this usually reflects a city government that is keen on filling in its urban policy and land-use planning according to more progressive-redistributive objectives aimed at providing decent and affordable social housing, as envisaged by broader objectives of the City Statute itself. In that sense, several cities of the ABC region have a reasonable track record in terms of reserving specific areas for ZEIS in their plans. This facilitated the relatively quick "fine-tuning" of Target Group 1 projects of MHML in reasonable locations in the city.[14]

For instance, this explains why in Diadema – a city with an extensive track record of participatory-collaborative master planning and slum upgrading – projects such as *Residencial Vitória, Yamagata, Condomínio da Gema, Ana Maria, Mazzaferro I* and *II* are all located within a perimeter of less than 2 kilometres from the city centre. As a matter of fact, it is hardly a coincidence that in this city all Target 1 projects are located in areas demarcated as ZEIS. Along the same lines, in Santo Andre, the projects *Residenciais Nova Conquista, Londrina, Juquiá* and *Guaratinguetá* (Estates 1–4) are also located in ZEIS, all within a perimeter of three kilometres from the urban centre. In Mauá, the project estates *Mauá 1* and *2* are three kilometres away from the central city, while the distance in *Altos Mauá 1* is five km. In São Bernardo, the projects *Residenciais Frei Betto* and *Nelson Mandela*, are also within ZEIS, located at less than two kilometres from the city centre. In any case, Figure 6.12 shows that all city centres are relatively close by, in the worst-case scenario, within a perimeter of seven kilometres.

Figures 6.13, 6.14 and 6.15 show that the majority of Target Group 1 projects and housing units were approved in Special Social Interest Zones, while only two Target 2 projects (*Ecovila and Portal da Vitória*) and one Target 3 undertaking

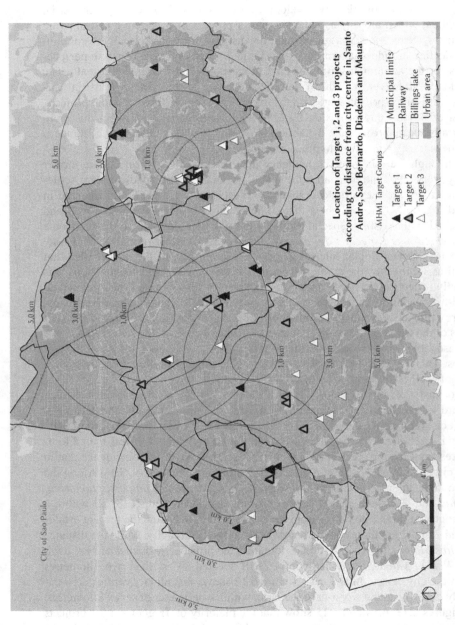

Figure 6.12 Location of projects for target groups 1, 2 and 3, according to distance from city centre in Santo André, São Bernardo, Diadema and Mauá

Source: Author's elaboration. Based on CEF (2015).

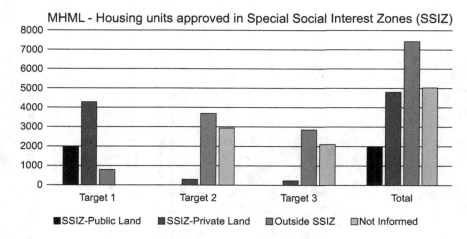

Figure 6.13 MHML housing units approved in special social interest zones
Source: Author's elaboration. Based on CEF (2015).

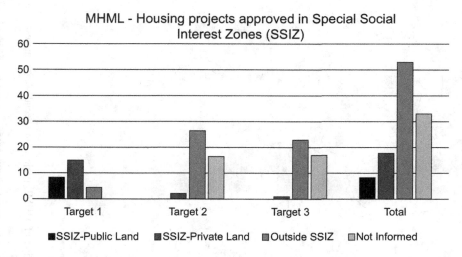

Figure 6.14 MHML housing projects approved in special social interest zones
Source: Author's elaboration. Based on CEF (2015).

(Viva Vista) were approved in these special zones in Diadema. In relation to Target 1 projects approved and located on public land, without exception, all of them were located in ZEIS, which indicates the match between broader urban policy making and the MHML program in the region.

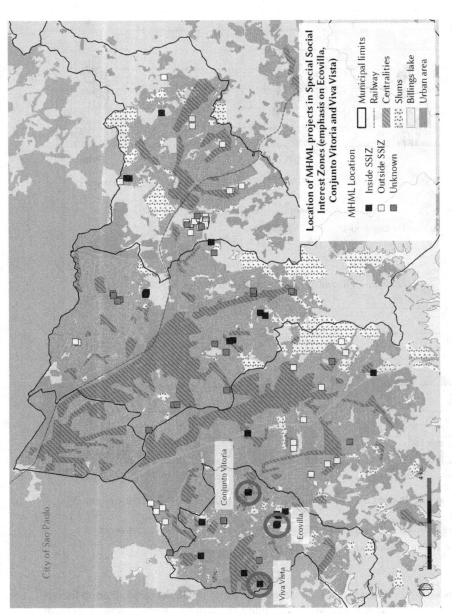

Figure 6.15 Location of MHML projects in special social interest zones (emphasis on Ecoville, Conjunto Vitoria and Viva Vista).

Source: Author's elaboration. Based on CEF (2015).

As mentioned, one of the critical arguments that emerged from the national review studies was that the program threatened to repeat the pattern of housing provision that had been prevailing under the NHB during the military regime by "producing massive, isolated housing estates without cities" (Ferreira, 2012). Figures 6.16–6.19, however, reveal that the majority of the projects have a dense network of infrastructure and community services within a range of 1 kilometre, such as bus terminals, libraries, kindergartens and primary education centres, cinemas, shopping centres and local commerce and hospitals, among others.

6.3.2.3 Building beyond the limits: On the hollowing out of MHML and subsidies flowing uphill[15]

Target Groups 2 and 3 projects are driven by the private sector. Therefore, an evaluation of MHML for these target groups requires a closer look at the projects and strategies of real estate actors and how they interacted with the program in the region. At the national level, from its launch until 2016 (excluding phase three), the MHML program had registered the development of 3.75 million units, out of which 2.05 million were contracted for Target Groups 2 and 3 (Pereira, 2019).

Our analysis undertaken on the basis of the specific data that were obtained from the regional office of the National Housing Bank (*Caixa Econômica Federal*) shows that part of the units that were originally contracted and designed for Target Groups 2 and 3 projects within the program eventually "trickled upward" and were commercialized outside MHML. More specifically, during 2009–2014, in the ABC region the program initially contracted 12,174 units for Target Groups 2 and 3, which were distributed over 86 projects. However, a closer look at the effective release and allocation of resources shows that only 6,148 units, in other words 50.50%, were effectively financed and distributed to the final beneficiaries of the program (Table 6.7). This was also confirmed by the reduction of credit that was allocated by the program in the region, particularly in 2013 and 2014 (Arcanjo, 2016: 68).So, what happened? Projects get effectively contracted after a detailed analysis by the implementing banks (either *CEF* or *Banco do Brasil*), particularly in relation to the guidelines of the program regarding price ceilings of the housing to be built, location and available infrastructure in the immediate surroundings. Once they get approved, Target Group 2 and 3 undertakings officially receive the label of "units/projects that have been contracted" by the program. This means that developers can use the official marketing of the program once they start building the units and, if they chose to do so, obtain construction finance from MHML. At this point, the program has officially contracted units, while potential buyers still have to be found.

Housing units "effectively financed by the Program" both comply with the general guidelines regarding price ceilings, technical characteristics and available infrastructure of the housing units, as well as the income limit of the final beneficiary that will take credit in order to finance their house. To be clear, housing units that are initially registered and contracted by the program but end up being bought with credit by beneficiaries that do not fall within the financial guidelines

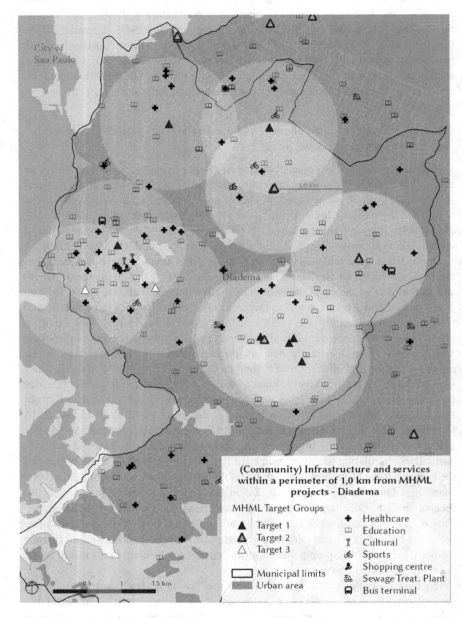

Figure 6.16 Community infrastructure and services located within a perimeter of 1 km from MHML projects in the city of Diadema

Source: Author's elaboration. Based on CEF (2015).

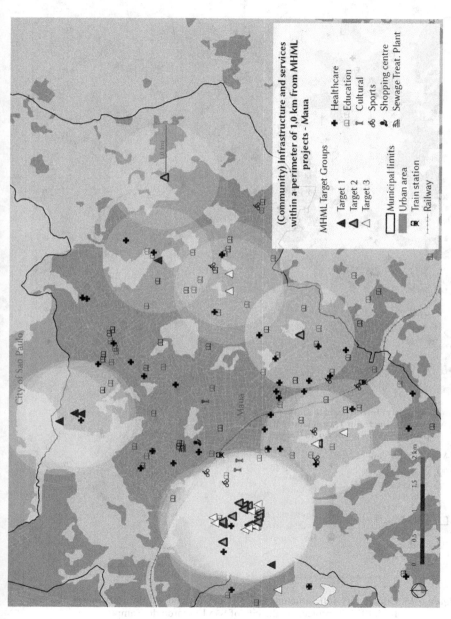

Figure 6.17 Community infrastructure and services located within a perimeter of 1 km from MHML projects in the city of Mauá

Source: Author's elaboration. Based on CEF (2015).

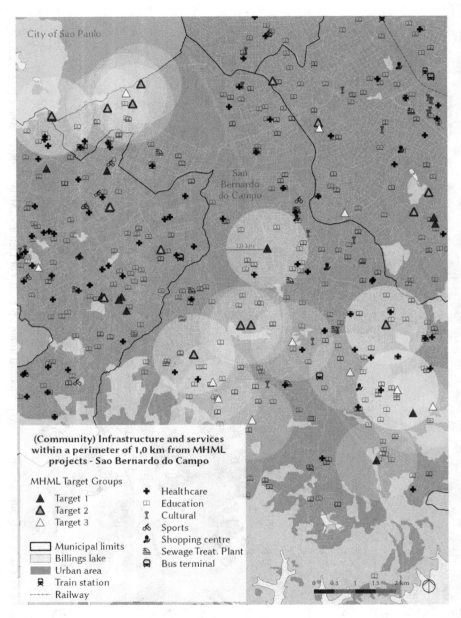

Figure 6.18 Community infrastructure and services located within a perimeter of 1 km from MHML projects in the city of São Bernardo do Campo

Source: Author's elaboration. Based on CEF (2015).

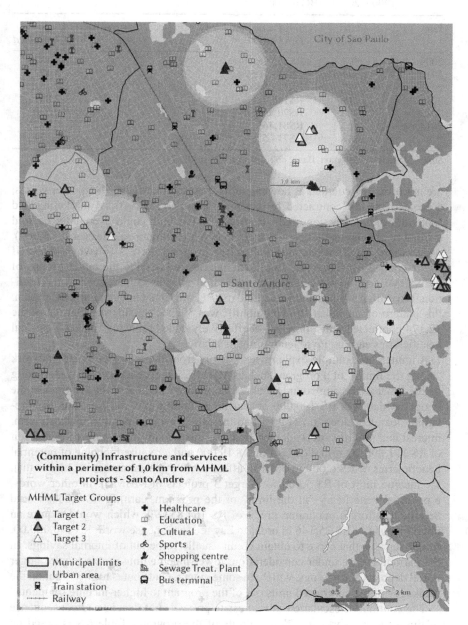

Legend (within the map):

(Community) Infrastructure and services within a perimeter of 1,0 km from MHML projects - Santo Andre

MHML Target Groups

▲ Target 1
△ Target 2
△ Target 3

▭ Municipal limits
▨ Urban area
🚉 Train station
┅ Railway

✚ Healthcare
▥ Education
⚲ Cultural
⚵ Sports
⚲ Shopping centre
⚙ Sewage Treat. Plant
🚌 Bus terminal

City of Sao Paulo

Santo André

1,0 km

0   0.5   1   1.5   2 km

Figure 6.19 Community infrastructure and services located within a perimeter of 1 km from MHML projects in the city of Santo André

Source: Author's elaboration. Based on CEF (2015).

Table 6.7 Target group 2 and 3 projects and units contracted and financed by MHML (2009–2014), ABC region

Year	Units financed	Total value financed (x R$ 1.000)	Projects Contracted	Units Contracted
2009	593	R$45,926	15	2,357
2010	1,532	R$135,105	25	3,472
2011	1,737	R$205,848	17	2,152
2012	1,219	R$166,426	17	2,801
2013	700	R$79,760	5	530
2014	367	R$81,465	7	862
	6,148	R$714,529	86	12,174

Source: Author's elaboration. Based on Klink et al. (2016: 72–73).

are not officially counted as housing units financed by MHML.[16] In other words, a project initially contracted by the program, using its official marketing and (potentially) construction credit from the FGTS fund, could eventually be sold to final beneficiaries that earned more than ten minimum salaries (or R$ 5,000), which represented the income limit of the program until 2016 (when phase III was launched). In this case, final beneficiaries are outside the target groups of the MHML and obtain credit from other, mid-income sources such as the traditional Brazilian savings and loans fund (*SBPE*, see Chapter 3). In the ABC region, this pattern of subsidies "flowing uphill" was 50.5%, considering that out of the 12,174 contracted units, only 6,148 effectively ended up being financed by final beneficiaries earning up to ten minimum salaries.

A numerical example explains how actors from the real estate markets work "along the limits" of a program such as MHML. Suppose a potential undertaking in 2014 that was being commercialized at the price ceiling of the program in the region, which was R$ 190,000 at that time. Taking into consideration the prevailing loan to value ratio of 90% and the usual financial parameters (loan term, interest and amortization) adopted by CEF, this would imply a maximum credit of R$ 171,000, an initial deposit of R$ 19,000 and a required income of potential beneficiaries of R$ 6,000 in order to afford monthly instalments (i.e. higher than the income limit of R$ 5,000 for Target 3 projects of MHML). In other words, a potential beneficiary "at the limit" of the program, earning R$ 5,000, would be able to afford a maximum credit of R$ 164,853.99, which would require an initial deposit of R$ 25,146 in order to buy the same house worth R$ 190.000,00. Initial deposits that serve to obtain credit usually come out of internal savings. In the particular case under consideration, lower income households require larger initial deposits and savings, which, to some extent, contributes to shed light on the "leakage" of the contracted units out of the program to higher-income segments.

The continuous escalation of real estate prices is both cause and consequence of this creative working with the limits of the program. Table 6.8 is based on specific data from the financial-economic and technical appraisal and supervision of projects, which were obtained from the Regional Office of the Housing Bank *CEF* (CEF, 2015). It can be seen that the average price of the units contracted

Table 6.8 Evolution of final unit prices, construction costs and land prices (per m²) in MHML projects in the ABC region, 2009–2014 (*)

Year	Nominal Values			Size Units	Real Values (Base Year: 2014)		
	Construction costs	Prices Units	Price Land	(m2)	Construction costs	Prices Units	Price Land
2009	R$807.48	R$1,999.90	R$391.07	57.66	R$1,158.12	R$2,732.88	R$534.40
2010	R$873.81	R$2,112.99	R$525.68	60.18	R$1,162.90	R$2,593.80	R$645.30
2011	R$963.67	R$2,885.43	R$471.42	49.66	R$1,193.23	R$3,370.13	R$550.61
2012	R$1,004.26	R$3,213.78	R$538.99	45.89	R$1,160.84	R$3,481.72	R$583.93
2013	R$1,029.24	R$3,422.44	R$507.43	54	R$1,100.67	R$3,513.48	R$520.93
2014	R$928.49	R$3,458.67	R$613.59	47.48	R$928.49	R$3,458.67	R$613.59
	14.99%	72.94%	56.90%	−17.66%	−19.83%	26.56%	14.82%

Source: Author's elaboration. Adapted from Klink et al. (2016: 76).

(*) Inflation correction for construction costs: According to INCC (National Index for Construction Costs/IBGE).

Inflation correction for land and final unit prices: IGPM (General Commodity Prices Index/FGV).

by the program in the region increased with 26.56% during 2009–2014.[17] At the same time, the size of housing units shrank with 17.66%. In other words, if we multiply the per square meter price of housing in 2014 with the size of units produced in 2009 and 2010, we would end up with a final sales prices of R$ 203,784.84, clearly beyond the price ceiling of R$ 190,000 that was established by the program.

The same table also shows how the specific real estate dynamics was unfolding while the program was being implemented in the region. As mentioned, in real terms, per square meter prices charged for finished units rose from R$ 2,732.88 in 2009 to R$ 3,458.67 in 2014 (i.e. an increase of 26.56%), while per square metre construction costs fell from R$ 1,158.12 to R$ 928.49 in the same period (that is, a decrease of almost 20%).[18] At the same time, the price of land per square metre increased from R$ 534.40 in 2009 to R$ 613.59 in 2014 (i.e. a growth of 14.82%).[19] Thus, while developers, builders and construction companies were using the argument of rising costs in order to demand higher price ceilings for housing units and more generous subsidies in their negotiations with the government about the required adjustments in the program, our data indicate that subsidies were actually capitalized and led to higher product and land prices, while per unit construction costs effectively decreased in light of the economies of scale that was enabled by the program. As a matter of fact, in her national review on the building and construction industry, Pereira (2019: 164) shows a similar pattern for the period 2007–2013: while the average price of housing units per square metre had increased with 77%, the average per square meter construction cost had fallen with 3%.

The prevailing macroeconomic scenario of renewed growth of income and employment generation, lower interest rates and allocation of additional liquidity through MHML contributed to accelerate this "drainage" of Target 2 and 3 projects to buyers who could finance their units under increasingly favourable conditions and, as such, were willing *and able* to contribute to a self-fulfilling prophecy of increasing prices (Mendonça and Sachsida, 2012). In this optimistic environment, it was actually rational for developers to simply opt to build outside the limits of the price and subsidy guidelines of MHML in order to increase their revenues and financial profits. While in times of macroeconomic crisis the large-scale volume of subsidies of MHML had performed well, both as a countercyclical instrument for income and employment generation and as a mechanism to revert investor's negative expectations, in this new scenario developers perceived the program had served its purpose and were increasingly moving up-market.

6.4 Conclusions and implications for future research

Looking at the Brazilian housing finance trajectory, it can be claimed that MHMV brought both innovation and continuity. The budgetary grants allocated to the poorest Target Group 1 beneficiaries represented a clear recognition that these segments were unable to access housing markets without adequate support from the public fund. In a way, although the volume of financial resources that were allocated represented a mismatch in relation to the size of the housing deficit that was by and large concentrated in Target Group 1, this design feature clearly echoed earlier discussions on the role of the state and the public fund in absorbing some of the risks associated with everyday life through a contribution to social wages for the poorest of the poor. At the same time, however, MHML signalled continuity; the main one being its emphasis on credit-driven provision of social housing, which was to be built by the private sector and financed by the traditional compulsory savings fund FGTS (see Chapter 3).

This partial continuity was probably also the reason that the first evaluations of the program raised the critique that, despite the emergence of a progressive national framework to guide housing policy and finance, including the dissemination of a new generation of participatory and redistributive master and land-use planning exercises, history seemed to be repeating itself. More specifically, Brazil was again building a massive volume of peripheral housing units without producing sustainable cities with more social justice. Moreover, the critics argued that the fast-track project-driven approach of the federal government was actually threatening to hollow out and bypass the broader project of urban and social reform itself, which had been carefully anchored at the local scale in the context of decentralization and democratization.

Our argument here was that the evaluation of a program such as MHML requires an approach that is sensitive to the geographical and historical specificities, contradictions as well as opportunities that emerge in the Brazilian structure of housing provision and finance. The analysis of the program's impacts on the ABC region showed that, although Target 1 units, built with budgetary grants for the poorest segments of MHML, were located outside the more dynamic and expensive segments of the real estate market, the areas selected were nevertheless far from peripheral. This was partly due to the fact that municipalities in the region had accumulated experience with progressive-redistributive planning instruments such as the ZEIS, which facilitated the match between good and relatively well-serviced locations and Target Group 1 projects, either on public or private land. Besides that, during decades, peripheral locations in the ABC region had been the object of transformative slum upgrading and infrastructure investments; as such, somewhat more peripheral locations were not synonymous to precariousness, as was reflected in the availability of community infrastructure and services in most of the Target 1 projects of the program within a perimeter of one kilometre.

At the same time, however, the implementation of the credit-driven Target 2 and 3 projects for the more affluent segments of the program were more cumbersome in the region. As a matter of fact, instead of MHML exclusively hollowing out urban reform, it could be argued that the knife cuts both ways. In other words, the limited progress of local governments in filling in a strategy aimed at combatting land and real estate speculation in general has generated incentives for developers to move up-market and build "outside the limits" established by price and subsidy guidelines of the program in order to maximize revenues and financial profits. In a way, then, the lack of leverage of local governments over land and real estate markets has led to a scenario of a significant hollowing out of the program's effectiveness when comparing contracted versus effectively financed units allocated to buyers earning up to ten minimum salaries.

At the time of writing this chapter, the innovative budgetary component that supported the operationalization of the grants for Target Group 1 projects has practically dried up in light of the economic crisis and the politics of austerity that have intensified after 2015. One more time, then, the poorest of the poor will carry the burden associated with the risk (financial, socio-environmental and so on) and the responsibility to provide and sustain their own housing in metropolitan areas. In the meantime, MHML is being redesigned in order to revisit the project of facilitating the penetration of finance into the built environment through mechanisms such as the right to buy (with credit) for Target Group 1 projects (instead of the prevailing outright transfer of housing units after ten years, under the condition of having paid heavily subsidized non-interest-bearing instalments of 5% of income), increasing price ceilings for housing units and decreasing loan to value ratios (that is, requiring higher initial deposits and beneficiaries' internal savings). In a dramatic way, this credit-driven approach with built-in mismatches between house prices, loan instalments and beneficiary incomes insists in moving into a policy direction where the results are pretty well known, at least in the Brazilian setting.

For instance, although in 2019 housing credit is again expected to increase, the five major Brazilian banks –*Banco do Brasil (BB), Itaú Unibanco, Bradesco, Santander e Caixa* – are reported to have accumulated around R$ 17 billion in real estate associated with loan defaults in the housing sector in light of the post-2015 economic crisis. Since then, banks have been consistently accumulating new real estate in light of loan defaults, but the figure peaked in 2018 with a growth of 32.3%. In the last two years (mid 2017-mid 2019), the compounded growth of this housing stock on the balance sheet of banks has been 78%. According to a recent diagnosis, the majority of this real estate is related to credit defaults of Target 2 and 3 projects of MHML.[20,21]

Notes

1 At the time of the creation of phase I of MHML (April 2009), this amounted to R$ 1,395.
2 Among others, a group of 11 research teams from different universities organized around the "Network on Cities and Housing" in order to design and implement a collaborative evaluation study of the MHML program, which was financed by the Ministry of Science and Technology and the Ministry of Cities. The teams and the specific regions analyzed were: the Federal University of Pará (metropolitan region of Belém and the South-eastern part of the state of Pará); Federal University of Ceará (metropolitan region of Fortaleza); Federal University of Rio Grande do Norte (metropolitan region of Natal); Federal University of Minas Gerais (metropolitan region of Belo Horizonte); Institute for Urban and Regional Research and Planning (IPPUR) and Post-Graduate Program for Urbanism (PROURB) from the Federal University of Rio de Janeiro (metropolitan region of Rio de Janeiro); the Institute POLIS (metropolitan São Paulo); the PUC University of São Paulo (metropolitan São Paulo with specific work on the city of Osasco); LABCIDADES (Faculty of Architecture and Urbanism, University o São Paulo) (metropolitan regions of São Paulo and Campinas); University of São Paulo in the city of São Carlos and the PEABIRU Institute (metropolitan region of São Paulo) and University of São Paulo in the city of São Carlos (Administrative regions of São Carlos and Ribeirão Preto). The final results of these collaborative research efforts were organized and published by Amore et al. (2015), using a provocative title: "My House . . . And the City? Evaluation of the Program My House My Life in six Brazilian States".
3 The abbreviation is derived from the first letters of the three economically more important cities, that is, Santo **A**ndré, São **B**ernardo do Campo and São **C**aetano do Sul, respectively.
4 To be specific, in March 2016, after the impeachment of President Dilma Roussef and already in the midst of an economic and political crisis, the federal government nevertheless managed to launch MHML-phase III, with considerably more modest delivery targets and limited implementation progress at the time of writing. See also Table 6.3 with the main summary statistics on Phases I to III of the program.
5 As shown in Table 6.3, however, phase III of MHML insists in repeating the classical pattern of Brazilian social housing finance by not *ex-ante* matching the available subsidies, target groups and the overall reduction of the housing deficit.
6 This section is based on Klink et al. (2016).
7 This study was based on data that were made available by the *Centro de Estudos da Metrópole* (CEM, meaning Centre for Studies on the Metropolis), which provided access to the real estate information system from the *Empresa Brasileira de Estudos do Patrimônio* (EMBRAESP, meaning the Brazilian Company on Real Estate Studies). For the specific purposes of providing a general overview of the real estate dynamics in the region we have concentrated on the period until 2011. The reason is that, after that year, the statistical series provided no longer allows an analysis according to the size and capital structure of developers.
8 We follow the classification of Hoyler (2014). Small developers: an individual yearly revenue base smaller than R$ 50 million and having only one or two projects/undertakings (representing 65.9% of the overall number of firms); Medium-sized

developers (making up 33.3% of the total number of establishments), with revenue levels in between R\$ 50 million and R\$ 2 billion; and big developers (0.8% of the overall number of companies), with developments between R\$ 2 and R\$ 23 billion.

9   Calculations were done on the basis of the General Price Index (IGP-DI) taken from the *Fundação Getúlio Vargas* (FGV). The annual geometric growth rate of this index during 2000–2010 was 9.64%, which was used to transform historical nominal values into present values of 2011.

10 It should be observed that the CEM-EMBRAESP database used here shows *launch prices* for new developments, which are not equivalent to the prices *effectively charged* for new developments.

11 To be specific, in many metropolitan regions (therefore including the ABC region), slum settlements are extremely dense. Therefore, upgrading operations frequently require partial relocation of the community and provision of alternative nearby housing units in order to guarantee the physical-technical-environmental sustainability of the remaining communities in slums. In the second phase of the program, the mayors succeeded in changing the operational guidelines of the program in order to facilitate the matching between slum upgrading and complementary provision of housing units.

12 All values were calculated in real prices in terms of the base year 2015.

13 In the last decade or so, this has been occurring in the older industrial corridors in São Bernardo do Campo (e.g. after the closure of the textile company *Tognato*), Santo André (in the *Tamanduateí* area around the São Paulo railway that connects São Paulo and the ABC region) and São Caetano do Sul (for instance, in the area of the *Cerâmica* neighbourhood). In the post-2006 scenario, the overheated real estate market increased the pressure to transform industrial areas to residential and mixed land uses.

14 As a matter of fact, before the creation of MHML, the private sector was rather reluctant to build and develop projects in ZEIS areas considering the perceived lack of financial feasibility. ZEIS were evaluated as a financial burden that required compensating subsidies. In all cities of the ABC region, the launch of MHML provided this compensation and triggered a significant development of social housing in ZEIS (Souza, 2018).

15 The following section is based on Arcanjo (2016) and Klink et al. (2016).

16 The same is true for the housing units bought without using credit, which is nevertheless a marginal phenomenon.

17 In present values of 2014 according to the general price index (IGPM).

18 In present values of 2014 according to the building construction index (INCC).

19 In present values of 2014 according to the general price index (IGPM).

20 The data on loan defaults and "subprime" housing on the balance sheets of banks were obtained from the financial newspaper *Valor*. Available at: www.valor.com.br/financas/6211435/bancos-nao-conseguem-dar-vazao-imoveis-retomados. (Accessed July 12th, 2019).

21 Likewise, Acolin et al. (2019) provide data on delinquency rates for Target Group 1 projects in six metropolitan areas (Belo Horizonte, Fortaleza, Rio de Janeiro, Salvador, Baixada Santista/Santos and São Paulo). In spite of the high level of subsidy for these target groups, they find that 28% of beneficiaries had not made a payment for 90 days or more as of December 2015. The authors explore four hypotheses for this relatively high rate of delinquency, that is, the peripheral location of units, insufficient income to cover ongoing user costs in housing estates (infrastructure, service costs and so on), deficient management of default and delinquency (generating issues of moral hazard) and the presence of organized crime. On the basis of these findings, they recommend a subsidy design that takes into considering the location of units (that is, higher subsidies for units in better locations in order to reduce delinquency rates). For the specific case of the ABC region, however, we have relativized issues of peripheral localization.

References

Acolin A; Hoek-Smit MC and Eloy CM (2019) High delinquency rates in Brazil's Minha Casa Minha Vida housing program: Possible causes and necessary reforms. *Habitat International* 83: 99–110.

Amore CS; Shimbo LZ and Rufino MBC (eds.) (2015) *Minha Casa . . . E A Cidade? Avaliação do Programa Minha Casa Minha Vida em seis estados Brasileiros.* Rio de Janeiro: Letra Capital.

Arcanjo RT (2016) *O Programa Minha Casa Minha Vida e o desenquadramento nas Faixas II e III: estudo de caso da região do ABC Paulista.* Dissertação de Mestrado. Programa em Planejamento e Gestão do Território, UFABC, Santo André.

Brasil (2013) *Programa Minha Casa Minha Vida.* Brasília: Ministério das Cidades.

Cardoso AL (ed.) (2013) *O Programa Minha Casa Minha Vida e seus Efeitos Territoriais.* Rio de Janeiro: Letra Capital.

CEF (Caixa Econômica Federal) (2015) *Balanço Físico-financeiro dos empreendimentos PMCMV na região do Grande ABC.* Santo André: CEF.

DiPasquale D and Wheaton W (1996) *Urban economics and real estate markets.* Upper Saddle River, NJ: Prentice Hall.

Denaldi R (2013) Trapped by the land? Change and continuity in the provision of social housing in Brazil. *International Journal of Urban Sustainable Development* 5(1): 40–53.

Eloy CM de M; Costa F de C and Rossetto R (2013) Subsídios na política habitacional brasileira: do BNH ao PMCMV. In: ANPUR (ed.) *Encontros Nacionais da ANPUR.* Recife: ANPUR, pp. 1–20.

Ferreira JSW (ed.) (2012) *Produzir casas ou construir cidades? Desafios para um Brasil urbano. Parâmetros de qualidade para a implementação de projetos habitacionais urbanos.* São Paulo: LABHAB and FUPAM.

Freitas FG; Whitehead C and Rosa JS (2015) *Finance and subsidy policies in Brazil and European union: A comparative analysis.* Brasilia, DF: MCidades; SNH; MPOG and Cities Alliance.

Hoek-Smit MC and Diamond DB (2003) Subsidizing housing finance. *Housing Finance International* 18(2): 3–13.

Hoyler TA (2014) Produção habitacional via mercado: quem produz, como e onde? In: Marques E (ed.) *São Paulo nos anos 2010. As transformações da metrópole.* São Paulo: Unesp.

Klink J and Denaldi R (2014) On financialization and state spatial fixes in Brazil: A geographical and historical interpretation of the housing program My House My Life. *Habitat International* 44: 220–226.

Klink J and Denaldi R (2015) On urban reform, rights and planning challenges in the Brazilian metropolis. *Planning Theory*: 1–16.

Klink J; Fonseca MLP de; Royer L; Feitosa F; Mello LF de; Sampaio L; Diniz J; Arcanjo RT; Alvarez G; Nogueira K; Sorrenti L and Peixoto T (2016) *Avaliação do Programa "Minha Casa Minha Vida" em São Paulo-SP e da Região do Grande ABC-SP.* Santo André: UFABC Relatório de Pesquisa. Chamada MCTI/CNPQ No 14/2013.

Krause C; Balbim R and Neto VCL (2013) *Minha Casa Minha Vida, Nosso Crescimento: Onde Fica a Política Habitacional?* Brasília: IPEA.

Magalhães I (2013) Planos Locais de Habitação na estratégia da Política Nacional de Habitação. In: Denaldi R (ed.) *Planejamento Habitacional: notas sobre precariedade e terra nos Planos Locais de Habitação.* São Paulo: Annablume, pp. 13–27.

Marques E and Rodrigues L (2013) O Programa Minha Casa Minha Vida na Metrópole Paulistana: Atendimento Habitacional e Padrões de Segregação. *Revista Brasileira de Estudos Urbanos e Regionais* 15(2): 159–177.

Mendonça MJ and Sachsida A (2012) *Existe Bolha no Mercado Imobiliário Brasileiro?* Rio de Janeiro: IPEA.

Pereira EC da (2019) *Preços Imobiliários e Ciclos Econômicos nos anos 2000: Uma Abordagem Heterodoxa.* Campinas: Universidade Estadual de Campinas, Dissertação de Mestrado em Desenvolvimento Econômico, Instituto de Economia.

Rodríguez-Pose A; Tomaney J and Klink J (2001) Local empowerment through economic restructuring in Brazil: The case of the Greater ABC region. *Geoforum* 32: 459–469.

Rolnik R et al. (2015) O programa Minha Casa Minha Vida nas regiões metropolitanas de São Paulo e Campinas: aspectos socioespaciais e segregação. *Cadernos Metrópole* 33: 127–154.

Royer L (2014) *Financeirização da Política Habitacional: Limites e Perspectivas.* São Paulo: Annablume.

Souza CVS de (2018) *Problematizando a reforma urbana no Brasil: Uma Abordagem de Escalas e Regimes de Organização e Intervenção Territorial do Estado.* Tese de Doutorado, Programa de Planejamento e Gestão do Território, Universidade Federal do ABC, São Bernardo do Campo.

Shimbo LZ (2012) *Habitação social de Mercado.* Belo Horizonte: Editora Arte.

UFABC (Universidade Federal do ABC) (2016) *Plano diretor regional.* Relatório final entregue ao Consórcio Intermunicipal Grande ABC, UFABC, Santo André.

World Bank (1993) *Housing: Enabling markets to work.* Washington: International Bank for Reconstruction and Development.

7   On reframing the capital market–austerity nexus and the emerging governance of urban securitization

7.1 Introduction

The global financial meltdown of 2008 has led to a proliferation of the literature on the linkages between the crises-driven restructuring of states, neoliberalization and austerity. The theme itself is not exactly new. To be specific, part of this literature echoes older debates on the reasons and mechanisms behind the gradual hollowing out of the financial governance that anchored the role of the public fund during the Fordist-welfare regime. Both in Western Europe and the US, the consistently growing fiscal deficits had raised doubts regarding the state's capacity to generate full employment and income through a continued reliance on the role of the public fund in maintaining the prevailing levels of accumulation, while providing contributions to the social wage and collective consumption (O'Connor, 1973; Oliveira, 1998). The post-1970 neoliberalization project had gradually provided a powerful representation of the ongoing productive and technological restructuring of the world economy and its crisis tendencies. According to that particular reading, the state and the public fund were both a cause and consequence of the crisis, which set the stage for subsequent calls for roll-back and retrenchment of governmental intervention. This particular ideological framing of the crisis concept, as well as its natural-organic relation with austerity and state retrenchment (Marcuse, 1981), has generated extensive debates since the demise of the Keynesian welfare regime.

But the subprime crisis has added complexity. For one, it has led to entanglements between the fiscal and financial dimensions of the crisis, marked by bail-out operations, state acquisition of toxic assets and similar emergency measures designed to save financial institutions "too big to fail". While credit was saved by money, and by exceptional monetary governance, the Italian, Greek and Spanish experiences have shown rather painfully that this has not come about without incurring significant costs in terms of increasing budgetary deficits, austerity measures and social deprivation.

The financial crisis of 2008 has also re-exposed the contradictory articulations between austerity and capital markets. According to the neoliberal project and narrative, cash-strapped governments require austerity in order to recover fiscal stability and responsibility, which are essential conditions to regain capacity to invest in new and maintain existing assets. This implies cutting expenditures, privatizing assets and bringing in know-how from private actors through a range of governance arrangements structured around public-private partnerships. Capital markets were always supposed to be key institutions to support (local) states in doing so, even if this meant bringing in the Trojan horse, as argued by authors such as Peck and Whiteside (2016). The latter's work on Detroit has clearly shown the dark side of the austerity-capital market nexus by analyzing the increasing penetration of norms and conventions of financial and budgetary responsibility in the making of post-political urban governance.

In the meantime, the austerity literature has evolved since the demise of Keynesianism, bringing in new ideas and themes from other disciplines beyond political economy, while generating comparative work on the variegated nature as well as the scalar dimension of austerity politics. For instance, an increasing number of authors have focussed on the framing and delimitation of crisis and austerity narratives, pointing to the strategic role of discursive strategies in the making and contestation of austerity, as well as in the design of counterhegemonic projects (Bayırbağ et al., 2017). Such an analytical perspective potentially provides useful insights to understand the surprising capacity of the neoliberal project to reinvent itself, particularly in the light of its contradictory socio-spatial outcomes and recurrent policy failures (Fuller and West, 2017).

Likewise, the emergence of what might be called a "scalar turn" in the literature has led to investigations on the strategic role of intergovernmental institutional and political relations within nation states, as well as the organizational, technical and financial capacity of cities in dealing with austerity (Keating, 2013; Adisson and Artioli, 2019). Cities are not mere receptacles of supranational entities and national states that unload austerity on the local level. Instead, urban austerity should be embedded in a fine-grained reading regarding the rescaling and restructuring of welfare regimes that is sensitive to geo-historical specificities and contingencies of cities and their social movements that strive to fill in alternative projects and trajectories.

Finally, crisis and austerity have increasingly been analyzed through the perspective of its (multi-scalar) impacts on urban and city-regional governance itself. This has generated a literature on a wide-ranging number of themes such as Peck's post-political austerity management and the imminent threat of state intervention in the setting of US cities (Peck, 2017a, 2017b), the role of European (new) social movements in contesting post-2008 austerity measures (in and against the state) in countries such as Spain and Greece and broader tendencies toward recentralization in order to guarantee fiscal-financial discipline and responsibility of cities and regional governments.

With a few exceptions, however, the specific role of finance and capital markets, as well as its entanglements with urban and city regional governance in the framing, managing and containing of the crisis-austerity nexus, has been surprisingly under-explored. To be clear, the implicit, ex-ante entrance point of much of the existing literature has been the disciplinary leverage of finance capital over urban management in a setting of austerity. As a result, most research has been undertaken on the investigation of the contradictory consequences of austerity in terms of the roll-back of social expenditures and its resulting spatial selectivity in cities, as well as the emergence of potential counterhegemonic projects and strategies driven by social movements, either supported by local governments or not. The borderline here is that finance and capital markets are generally framed as a "problem" in the planning and management of urban austerity. Nevertheless, this tends to underestimate representations, or *fantasmatic* logics in the terminology of Fuller and West (2017), whereby finance capital is the "solution" (or false solution for that matter) to austerity in light of the cash flows and professional know-how it is supposed to generate. Particularly in institutional settings whereby, for some reason or another, access to borrowing has either been constrained or limited in light of the lack of deep and consolidated capital markets, the axiom of finance and capital markets as a "(false) solution" becomes instrumental in a research agenda structured around the critical investigation of austerity urbanism in times of emerging financialization.

This brings us to a related silence in the contemporary literature (i.e. on austerity and finance in The Urban Global South). Most discussions have privileged the impact of crisis-driven state restructuring in the core, or, at best, in peripheral countries (such as Spain, Greece and Ireland) of the Fordist-welfare regime, while the investigation of the Global South has, by and large, not gone beyond more generic work on subordinated insertion within the international monetary governance and its contradictory effects on national space economies. The Brazilian case is emblematic.

There is significant Brazilian literature that relates neoliberalization, the rolling back of the national developmental state and the (external and internal) debt crisis with austerity governance (Affonso and Silva, 1995; Fiori, 1999; Tavares and Fiori, 1997). There is less work, however, on how fiscal austerity narratives and policies have evolved more recently, marked by the re-emerging role of finance capital in general and its entanglements with cash strapped cities, metropolitan areas and states in particular.

To be clear, during much of the 1990s, Brazilian local and state governments were subjected to harsh national austerity measures without compensating redistributional policies. The earlier mentioned Plano Real imposed deflationary macroeconomic stabilization and inflation control, which was accompanied by a package of austerity measures aimed at curbing expenditures of states and cities, selective federal incentives to the privatization of public banks and infrastructure as well as wide ranging restrictions on sub-national borrowing, while recentralizing part of discretionary non-tax sources of income (Klink, 2001).

The post-2013 scenario of economic crisis and stagnation of the income base of local and state governments has put austerity politics back on the agenda, but in a macro-institutional and political setting that is somewhat different from the one in the 1990s. A key element is the approximation between subnational states and capital markets. An increasing number of states and cities have been creatively looking for new institutional and financial devices aimed at accessing capital markets, even in a national institutional framework that puts significant restrictions on doing so since the previous austerity regulation of the 1990s. As will be discussed along this chapter, city and state governments have invented their own, rather opaque and non-transparent praxis in linking in with capital markets, which has generated a series of financial contradictions as well as contestations regarding the (il)legality of such operations. At the same time, however, the circulation of norms, conventions and devices regarding these "not-so-good" practices has influenced discussions in federal parliament regarding the approval of a new legal framework on securitization of tax receivables and other state assets. If the proposal passes, it will effectively legalize and legitimize the ongoing local praxis and set the stage for the consolidation of a national market for the securitization of state assets at multiple scales.

The discussion in this chapter provides three contributions to the existing literature on urban austerity, neoliberalization and financialization. First, it not only recognizes the multi-level character of austerity, thereby moving away from a stereotype representation of cities as receptacles of national states that download financial burdens but provides a relational scalar perspective whereby national off-loading is combined with local, potentially subversive and innovative experimentation with devices that enable borrowing and access to capital markets as the (false) solution to austerity. In a subsequent stage, this creative but contradictory praxis is circulated and potentially up-scaled in order to fill in a full-fledged national framework aimed at the securitization of state assets.

Second, the analysis developed in this chapter complements the prevailing literature on austerity and urban governance in industrialized countries, whereby the framing of finance and consolidated capital markets as a problem represents an axiomatic entrance point of much of the investigations. In the Brazilian setting marked by emerging institutions, however, finance and capital markets are framed as potential solutions to austerity by tapping into the available savings and professional know-how. Rather than prioritizing the analysis of the contradictory impacts of full-fledged capital markets on urban governance, as well as its contestation by social movements, such an approach requires fleshing out the role of the state, at multiple scales, in constituting the relationship with thin and risk-averse capital markets as "a solution", (or a false solution) in the first place. This involves investigating the financial and institutional devices, as well as the discursive strategies that are used to legitimize the supposed "intrinsic value added" of establishing closer linkages between cities and capital markets. The latter are packaged around narratives of a streamlined articulation between savings and investments, particularly in light of the historical absence of domestic capital markets in Brazil (Chapter 3). Thus, our approach in this chapter sheds light on the almost "performative" dissemination and circulation of a series of local "not-so-good" practices in constituting a city-capital market nexus, in spite of the lack of institutional transparency and financial sustainability that have characterized this recent local experimentation.

Third, the chapter shows how cities and states (and not national government) have effectively been first movers and innovators in the sense of exploring "the playing field" as well as stretching the legal and financial limits aimed at reducing systemic risks for investors in building the connections with markets for debentures and new products for securitization. Local praxis has gone beyond the usual stories on securitization of real estate or mortgage-backed assets and penetrates into the capitalization and securitization of a diversified flow of general-purpose tax and non-tax resources, without being restricted to a specific territorial perimeter, such as in the case of US tax increment finance districts or Brazilian Urban Partnership Operations with securitized building certificates (see Chapter 4). Moreover, local experimentation and "innovation" has frequently been backed up by a series of specific guarantees related to land and infrastructure assets. In a way, we will argue that the variegated "urbanization of finance" is instrumental in the securitization of urban space itself.

After this introduction, the chapter is organized in three sections. The first provides a brief overview of the literature on crisis, austerity and the restructuring of the nexus between states and capital markets. The second section is divided in two parts. The first summarizes the playing field of Brazilian fiscal federalism, prioritizing the analysis of the restrictions on subnational borrowing that has been laid out since the first generation of austerity measures in the 1990s. The second part provides key empirical evidence of how city and state governments have constituted an innovative, albeit subversive, subnational praxis that has increasingly bypassed national restrictions on borrowing and securitization. The conclusion wraps up the empirical evidence and provides elements for a research agenda on the rescaling and restructuring of states spaces, capital markets and austerity, with a particular relevance for countries in the Global South.

7.2 Crisis, urban austerity and governance

7.2.1 Introduction

In order to set the stage for the subsequent empirical discussion, this section provides a brief introduction to the literature on crisis, urban austerity and governance. As mentioned, since the earlier debates of the late 1970s, the research agenda has gone through an analytical sophistication and opened up to complementary disciplinary perspectives from wide-ranging areas such as political economy, sociology and discourse theory. Moreover, the literature has generated a series of empirical studies that have pointed out the variegated nature of austerity politics in specific geographic and historical trajectories, which also enabled to revise and enrich the broader claims of theories.

In what follows, we provide a preliminary attempt to classify this literature according to three broad complementary strands of work. A first approach, which is influenced by discourse theory and analytical perspectives sensitive to issues of subjectivity, is concerned with the framing, as well as the contested filling in of the crisis concept itself, considering the insights this generates for understanding austerity and the transformations in urban governance. A second prioritizes the investigation of the relational-scalar nature of austerity, whereby cities do not merely represent receptacles of downloading tactics from national and supra-national institutions, but arenas of ongoing struggles regarding the design, direction and distribution of austerity, or, for that matter, anti-austerity politics. A third strand in the literature is focussed on understanding the consequences of austerity on urban governance and planning, including the potential role of civil society actors and social movements in the contestation and redesign of austerity urbanism.

7.2.2 Whose crisis is it anyway?

A first strand in the literature is concerned with an interdisciplinary opening up of the black box concept of crisis itself, which for (too) long has been the domain of dogmatic orthodox economics. The latter has tended to normalize a crisis as an "organic" dimension of capitalism (i.e. as part of a broader productive and techno-logic restructuring of economies that requires a "natural" downsizing of the state).

As a matter of fact, this line of work goes a long way back when we consider Marcuse's (1981) analysis on the "manufactured" character of the urban fiscal crisis in US inner cities at the beginning of the 1980s. To be specific, Marcuse claimed that the mainstream argument regarding the inevitable decadence of large US cities, as well as the uselessness of resistance against the policies proposed, represented a "fraudulent" crisis (p. 342), and was effectively providing the legiti-macy for austerity measures targeted at rolling back social protection and housing subsidies for inner-city areas. An alternative framing would recognize the active role of the state itself, at multiple scales, in producing the inner-city fiscal crisis, for example, through the massive allocation of public funds in order to stimulate private investments, housing and (road) infrastructure in the suburbs, while leav-ing the poor and the black in the spatial trap of central-city areas.

Recent approaches have built on these earlier efforts. Bayırbağ et al. (2017: 2026–2031), for example, analyze a variety of urban crisis framings and prob-lematics. They develop an analytical lens that incorporates six complementary dimensions of crisis framing, that is, "Structure" (the political economic restruc-turing, the socio-spatial contradictions and the emerging "fault-lines"); "Alien-ation" (i.e. how contradictions are contained or result in a breach and are worked out in fully articulated urban crises); "Politics" (i.e. who makes and owns the hegemonic crisis narrative?); "Construction" (i.e., emphasis on process and sub-jectivity, involving meaning-making through discourse and identificatory prac-tices); "Boundaries" (i.e. crisis delimitation, involving the limits and potentials of re-establishing equilibrium); and "Indeterminacy" (i.e. the tipping point of crises, which reflects the tensions between rupture and recuperation).

Fuller and West (2017), on the basis of Foucault and post-structural discourse theory, further develop an approach that increases the understanding of the entan-glements between individual subjectivities, narratives and the shared universe of meanings, as structured by certain social, political and *fantasmatic* logics. The latter relates to "both foretelling of disaster and guaranteeing future harmony, demonizing certain groups and practices and approving of others" (p. 2092). This perspective becomes particularly promising in light of the lack of effective contestation of austerity itself. In other words, it provides potential insights into understanding how austerity is performed, how it can assume an epistemological function (Oosterlynck and Gonzales, 2013) and how it persists, in spite of its increasingly contradictory and socio-spatially selective tendencies, and, moreover, "the ways in which practical policy failure can serve to bolster ideological success" (Fuller and West, 2017: 2091).

7.2.3 State rescaling and austerity

A second perspective in the literature focuses on the scalar dimension of austerity as related to the restructuring of state spatial projects (the territorial organization of the state) (e.g. through centralized-decentralized arrangements; homogenous-customized mechanisms and so on) and strategies (the way the state is effectively involved in the production of space, for example, through regulation and investment) (Brenner, 2004). Adisson and Artioli (2019), for instance, argue that austerity is not simply offloaded to the local level, considering that in many European welfare systems built-in mechanisms of multi-level governance (Keating, 2013) have frequently allowed for a redistributive softening of austerity measures. Likewise, the uneven distribution of institutional, managerial and financial capacity at the city level has accounted for considerable local and regional variation in austerity policies. As a result, they derive a classification of four regimes according to the degree of intergovernmental checks and balances that are able to mitigate the impact of fiscal crisis on cities, on the one hand, and the institutional, technical and financial capacity of local governments to deal with austerity, on the other. To be specific, in their terminology a strong re-distributional intergovernmental system with low local capacity leads to "nationally mitigated austerity" regimes, while national institutions that tend to offload financial responsibilities to city government with strong capacities generates "locally mitigated austerity". "Gridlock austerity" occurs when national government downloads to cities with weak capacities, while "opportunistic austerity regimes" emerge in a setting marked by the significant presence of intergovernmental redistributive mechanisms and local governments with financial-institutional capacities to deal creatively with austerity.

This analytical perspective has also been put to work in order to establish an empirical dialogue with the ongoing process of restructuring and rescaling of austerity politics in specific places, taking into consideration their geographic and historical trajectories. Martí-Costa and Tomàs (2017), for example, focus on the dramatic rise and fall of the Spanish real estate and finance-driven miracle since the 1970s. The Spanish experience after democratization and partial decentralization of the 1980s (i.e. to the regional level of the *Comunidades Autónomas*) has followed a somewhat different periodization and is more hybrid than suggested by the framework adopted by Brenner (2004) for the core countries of European spatial Keynesianism. For instance, they show that during the rise of its real estate tourism and finance-driven growth regime (approximately between 1992 and 2007), the competitive rescaling to the regional *Comunidades Autónomas* through the massive rollout of speculative entrepreneurial redevelopment projects that were aggressively financed by the regional saving banks, was combined with a variegated expansion of welfare services in areas such as health, education and urban infrastructure. The Spanish post-2007/2008 scenario has triggered new rescaling processes, with an increasingly strategic role of the European troika in putting priority on the preservation of fiscal stability in order to guarantee payments of debts and interests. Constitutional reform and recentralization have consolidated a role for the national scale state as the "gatekeeper" of austerity guidelines from the European community, while offloading responsibilities for budget cuts in health, education and housing/infrastructure to the regional level of the *Comunidades Autónomas*. In a way, the Spanish scenario suggests that austerity regimes are drivers of organizational and institutional change (Martí-Costa and Tomàs, 2017: 2118). At the same time, however, social movements have clearly re-emerged in the scalar contestation of austerity. Somewhat different from the demands for democratization and the complaints about the effects of economic crisis, which characterized its previous agenda, the strategies of social movements have now shifted to the contestation of financialization and neoliberalization that have encapsulated Spanish cities and metropolitan areas. Movements have also connected with left-wing parties (PODEMOS) and are associated with recent electoral victories, both at the local and European level.

The filling in or hollowing out of particular scales is not neutral in terms of the politics of austerity. In relation to the Italian setting, for example, Armondi (2017) analyzes the recent changes in the regulatory framework that guides the planning, management and finance of Italian cities and regions through the creation of a national law on new metropolitan cities (law n. 56/2014). In the author's view, the involvement of the European and national government in the creation of this institutional framework represents a clear centralization of policy trajectories as well as a technocratic disruption of existing mechanisms of urban and regional governance, legitimized through generic notions of economics of scale and reduction of inefficiency. More importantly, it indicates a tendency whereby rescaling has increasingly become "a tool for austerity" (Armondi, 2017: 178).

Although not much work along this perspective has been undertaken for the Global Urban South, Alke (2018) analyzes authoritarian neoliberal rescaling through the lens of austerity urbanism in the intermediate Mexican city of Oaxaca, examining the increasing problems of security.

7.2.4 Austerity urbanism and governance from the inside out

A third line of research looks at the specific direct and indirect impacts of austerity and austerity narratives on urban governance and planning. Initial efforts have been undertaken around the earlier-mentioned emblematic case study cities such as Detroit, Atlanta (Peck, 2017a, 2017b) and Athens (Arampatzi, 2017). This empirical work has been expanded to include less extreme, "ordinary" experiences of urban governance in times of austerity and link it with efforts to see broader patterns in individual experiences in order to advance in the elaboration of broader theory.

For instance, on the basis of a detailed examination of four North American cities (including, in addition to Detroit, the cities of Dallas, Philadelphia and San Jose), Hinkley (2017) describes how governance in cities, as sites of variegated neoliberal localization (Brenner et al., 2010) both contain and circulate local crisis narratives and relate these to the broader context of solutions (or, for that matter, false solutions) that are circulating. While previous US crisis narratives were related to excessive social spending and unaffordable expenditures associated with the welfare state, the more recent representations and framings have focussed on "inefficient local governance, unsustainable pensions plans and un-going [sic] economic precarity" (Hinkley, 2017: 2123). In such a context, crisis and the proposed mechanism to solve it, become potentially contagious, considering that "comparison treats multiple cases as sites of participation in shared processes, which are constituted in and through that variation" (Hinkley, 2017: 2125). In the particular set of cities studied, the discursive strategies that emerged to solve crisis tendencies emphasized reform in supposedly unviable municipal pension plans as well as state intervention in inefficient municipal fiscal management or city governance that was unable to take the necessarily tough (post)-political decisions. In relation to pensions, in cities such as Detroit this particular solution was rather misplaced considering the demographic and economic drying up of the city. Moreover, in the other cities the pension narrative allowed to move away from other possible strategies such as revenue-raising options or a broader national reform in health care and retirement schemes. In relation to state intervention, it should be recalled that US states, de jure, have considerable potential influence over cities in case of local financial mismanagement or threats of bankruptcy. This context has implications on how to interpret the narratives of crisis that emerge from cities. For instance, receivership laws potentially remove fiscal decision-making from the local scale and provide powers that elected city officials usually don't have, such as the capacity to dissolve contracts and enter into collective bargaining agreements. In extreme experiences such as Detroit, emergence managers try to avoid bankruptcy, considering that the claims of pensioners would potentially be given prevalence over those of bondholders (Hinkley, 2017: 2133). State intervention usually works with sticks instead of carrots. This implies cutting expenditures, selling assets, breaking union agreements, restricting local borrowing and reducing intergovernmental state aid, among some of the more frequently adopted strategies. Moreover, longstanding federal policies of withdrawal of intergovernmental transfers have been aggravated by a deliberate, discretionary reduction of the flow of funds from states to cities. Thus, according to the authors, narratives on pensions and state intervention have become part of a broader process of "financialization of urban policy and governance itself", giving privilege to technical instead of political solutions.

Hall and Jonas (2014) provide one of the few examples in the urban austerity literature that explores in a more detailed way the entanglements between city governance and the capital market itself.[1] More particularly, their example of metropolitan specific-purpose governance for transportation in Detroit is intriguing, considering the dramatic financial scenario of its core city. In line with the overall trajectory of the American metropolis, Detroit has a long-standing history of failed attempts to create collaborative governance arrangements for transportation. Part of this failure is due to deep racial and socio-spatial disparities, as well as the institutional fragmentation that has separated the inner city and its more affluent suburbs. More recently, however, the creation of a Regional Transportation Authority (RTA) has allowed what the authors call "a dual spatial fix" (Hall and Jones, 2014: 189) in the sense that the new city-regional/metropolitan design enabled the core city and the suburban municipalities – also affected by a shrinking revenue base in light of the global financial crisis – to access capital markets jointly and effectively bypass restrictions on borrowing, while providing developers and members of the local growth machine opportunities to project increased rents and financial profits through more intense land uses along the corridor of the proposed infrastructure network. At the same time, however, the authors conclude that the complex financial-institutional engineering shows a tendency to produce relatively opaque specific-purpose arrangements that separate a diversified flow of future tax revenues from the general-purpose municipal budget, generating increasing financial risks and lack of transparency in governance.

In the next section we will further flesh out these entanglements between finance and capital markets, governance and austerity urbanism for the specific Brazilian trajectory.

7.3 Capital markets and austerity urbanism: From a problem to a false solution?

7.3.1 Rolling back the developmental state and downscaling fiscal responsibility

The fiscal and financial design behind the macroeconomic stabilization and inflation control of the Plano Real in the 1990s envisaged rescaling as a key tool to achieve austerity. To be clear, in spite of the downsizing and outright privatization of federal institutions that had taken place since the 1980s, the architects of the Plano Real perceived the presence and activism of state-owned banks, and the consequent ease of access to borrowing of cities and states as a clear legacy of the developmental state which, if untouched, would threaten to undermine the long-term success of the plan. According to that view, subnational fiscal indiscipline and the political use of state-owned banks had generated a pattern of recurrent bailout operations and increasing federal deficits. Business as usual would surely challenge the stabilization of inflation. The broader international policy community subscribed to this view. For instance, in circles of the IMF and the World Bank, any financial support to the country's troubled financial sector would be conditioned to the extinction or privatization of state banks in order to avoid monetization and escalation of price levels (Costa, 2012). Thus, downscaling austerity politics and putting restrictions to subnational borrowing were perceived as pillars to sustain the initial success of the real.

In that sense, the launch in 1995 of the so-called Support Program for the Structural and Fiscal Adjustment of States kick-started a series of federal initiatives, with carrots and sticks, framed around the renegotiation, or refinancing of subnational debts under the condition of downsizing personnel, privatization or concessions of infrastructure services, state banks and companies, modernization of tax and revenue raising, achievement of specific fiscal targets and strict control and supervision of local borrowing.[2] In the meantime, inflation control had reduced the financial profits obtained from the overnight market and short-term floating of deposits, which had exposed the structural deficiencies of state banks. As a result, the federal government seized the opportunity to obtain more leverage in its negotiations with states. The 1995 program was subsequently extended to local governments through Law N° 9.496 (September 11, 1997) and decree N° 2.118, and by then had already mobilized a significant amount of financial resources (more particularly R$ 103 billion), including the amounts spent on the financial restructuring of state banks (Nascimento and Debus, 2001). These authors estimate that structural adjustment involving the conditional "clean-up" of balance sheets and renegotiation of municipal and state debts was responsible for almost half of the increase of the stock of debt of federal government between July 1994 and December 2001.[3]

Negotiating and downloading austerity politics implied financial centralization. The conditions imposed on São Paulo, the richest state of the federation, were emblematic. On May 20, 1997, it finalized a loan agreement of R$ 50.4 billion that ultimately enabled the privatization of the state-owned bank *Banespa*, which would be acquired by the Spanish group Santander. At the time it represented the biggest debt deal in the history of the country, considering that the renegotiation of the Brazilian debt with its international creditors, which occurred in the first half of the 1990s, was approximately 40 billion USD.[4]

In this context, the approval of the constitutional Law N° 101 (from May 4, 2000) on "Fiscal Responsibility" represented something of a cherry on top. It was influenced by the circulation of ideas that had underpinned the IMF's criteria on transparency and accountability in budgetary planning and management, as well as the mechanisms used in the setting, monitoring and supervision of targets for fiscal deficits and debt levels in the context of the negotiations of the Maastricht Treaty in Europe, among other examples. Its main objective was to consolidate more transparent and financially sustainable fiscal planning and management, which would generate a trajectory of gradual reduction in the stock of subnational debt towards specified targets. As a matter of fact, the *leitmotiv* behind the law was to supervise cities' and states' access to capital markets and control local borrowing in order to avoid the excessive accumulation of debts that had marked developmental governance in recent history. Primary surplus, meaning the budgetary surplus excluding capital market operations, became a key metric.

The subsequent senate resolution N° 40 of 2001 fleshed out the National Law on Fiscal Responsibility in terms of a series of operational guidelines and requirements, such as the gradual reduction of the ratio of the stock of total debt to net operating revenues for federal, state and city government to target levels of 3.5, 2.0 and 1.2, respectively, considering a time span of 15 years. Before reaching these targets, access to capital markets would be effectively blocked. Likewise, this senate resolution provided restrictions regarding the cash flows involved in the amortization and payment of interest, which couldn't exceed 11.5% of net operating revenues of subnational bodies. The legislation adopted a broad approach in terms of defining its objects and outreach; credit operations included the emission of titles, mercantile leasing and the use of a range of derivatives and mechanisms to anticipate expected flows of revenues. Moreover, in addition to the institutions that are part of direct administration, the law included "dependent state enterprises", meaning companies that were controlled by states or municipalities and received resources in order to finance their operating or capital expenditures. Nevertheless, in relation to the latter, a lump-sum inflow of resources associated with an increase in the participation of shares in controlled companies, authorized and approved by the local legislature, was not considered as financial dependency.[5] Since 2002, non-compliance with the targets established for debt reduction and liquidity ratios in the national law have represented effective threats, considering the severe penalties that are applied, such as the blocking of intergovernmental financial transfers and the potential impeachment of local officials, among other examples. Finally, local borrowing operations have been supervised by the Ministry of Economic Planning according to the earlier-mentioned guidelines of the senate (that follow the National Law on Fiscal Responsibility), and also require approval from the municipal or state legislature.

What has been the impact of austerity governance, as promulgated by Law N° 101, in terms of its effective performance to reduce cities' and states' appetite to borrow? Figure 7.1 shows the evolution of national, state and municipal debt

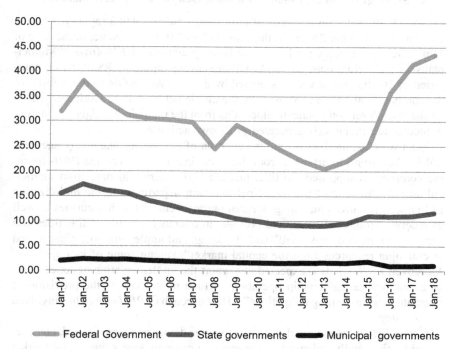

Figure 7.1 Net public debt as a % of GDP (Dec. 2001–Dec. 2018): Federal, state and municipal government

Source: Author's elaboration. Based on Banco Central do Brasil.

levels as a percentage of GDP since 2001. Leaving aside the abrupt increase in the figure from 2001 to 2002, which can be attributed to factors such as reduced macroeconomic growth and the electoral climate that marked the year 2002, Law Nº 101 has delivered up its promise in terms of providing a consistent reduction of debt as a percentage of GDP until December 2013, while also successfully mobilizing states and local governments in an effort towards sustaining balanced budgets. Things change, however, after the deterioration of the political and economic climate in 2013. Federal government, after having successfully centralized the interface with capital markets during the 1990s, has been the first to aggressively reconnect with finance as a solution for austerity; net federal debt as a percentage of GDP increased from 20.45% in 2013 to 43.38% in 2018. Evidently, part but not all of this is due to the macroeconomic stagnation that has marked the country since 2014. At the same time, local and state governments have been forced to face austerity and depressed income and employment levels in the more restricted regulatory environment that guides access to capital markets since the Law on Fiscal Responsibility. As will be discussed in the next section, this has now triggered their institutional and financial creativity in order to re-connect with finance capital.

7.3.2 Urban governance, finance and the search for solutions to austerity[6]

On March 29, 2011, on the basis of municipal Law Nº 10.003 (approved by the council on November 25, 2010) the capital city of Belo Horizonte, in the state of Minas Gerais, created a public limited liability company *PBH-Ativos* ("Municipality of Belo Horizonte-Assets"), with a starting capital of R$ 100,000, composed of 10,000 ordinary shares owned by the city. According to the website of the company, "although controlled by the municipality of Belo Horizonte, *PBH-Ativos* doesn't rely on financial allocations from the municipal budget. Its sources of income are exclusively generated by its own activities"[7]

What were the motives behind the creation of a company like that (Canettieri, 2017)? According to article 2 from the municipal law that created *PBH-Ativos*, the company is supposed "to title, manage and generate income from municipal assets", "to support the municipal planning department in the mobilization of financial resources including, for that purpose, emit its own securities as well as receive, acquire, alienate, and provide, in guaranty, assets, credits, titles and shares from the company itself' and "to design and implement operations aimed at obtaining resources from the capital market", among some of its objectives. Moreover, *PBH-Ativos* is expected to support the implementation of concessions or public-private partnership operations, which may involve "the provision of guarantees and assuming obligations".[8] The approach of *PBH-Ativos* has involved various steps.

First, *PBH-Ativos* emerged as a rather opaque device as a state-owned, but, presumably, financially independent company in order to access the capital market and bypass the earlier-mentioned National law on Fiscal Responsibility. *PBH-Ativos*, as a state-owned securitization company, served as a legal-financial device in the use of public funds and urban assets as guarantees to access the private market of debentures. More specifically, *PBH-Ativos* mobilized its municipal real estate and receivables linked to infrastructure income and taxes (the underlying base being delinquent taxes or taxes in arrear) as guarantees in its operations with municipal debentures. For instance, on the January 10, 2014, the municipality approved a law (N° 10.699), which enabled it to transfer to *PBH-Ativos* a stock of real estate and land assets with a net worth of R$ 154,956,671. In relation to infrastructure receivables, the municipality mobilized a resource base that had been created as part of a broader agreement with the state government of Minas Gerais regarding the shared maintenance and investment in environmental sanitation in the city of Belo Horizonte. The latter required the state sanitation utility COPASA to deposit, until December 2031, monthly instalments into a municipal fund. In April 2011, the net present value of these future instalments from the state represented an asset of approximately R$ 245 million, which was transferred by the municipality to *PBH-Ativos* in order to increase its capital (Fattorelli and Gomes, 2018). Finally, in order to mobilize the municipal tax receivables that were linked to delinquent taxes or taxes in arrear, the financial engineering that was designed involved the use of subordinate and ordinary debentures in an articulation between the city and the capital market that was mediated by *PBH-Ativos*. To be specific, the subordinate debentures issued by *PBH-Ativos* were nothing more than a legal device that was used in order to enable the municipality to transfer its tax receivables to the securitization company in a specific separate account. In exchange, the municipality received non-tradable, non-interest paying titles that were anchored to the same monetary correction and legal penalties for delinquent taxes and taxes in arrears, as laid out in the tax code that guided the municipal receivables. The ordinary debentures issued by *PBH-Ativos*, however, represented an effective capital market emission, which used the specific account of the tax receivables that had been created as a guarantee. Ordinary debentures received monthly interest at prevailing market rates and under conditions for monetary correction and penalties for delay that were specified in the contract. For instance, in 2014 *PBH-Ativos* ended up paying a real rate of 11%, implying a nominal interest of 23% (CNACD, n.d.). Figure 7.2 summarizes the financial-institutional device that was behind *PBH-Ativos* and which would serve as a source of inspiration for other similar experiences of state-mediated securitization.

Second, the constitution of state-owned securitization companies like *PBH-Ativos* represented something of a camouflage of the effective contradictory risk-sharing that was at stake in this type of operation with municipal debentures. On the one hand, the local government transferred its tax receivables in a separate account in exchange for non-interest paying subordinated debentures. Simultaneously, however, the city remained fully responsible for the planning and management of tax receivables, as well as assuming additional obligations to provide all the necessary information to investors regarding the effective and projected flow of tax revenues according to the criteria of transparency and accountability that guide good governance in the capital market. On the other hand, the ordinary debentures with public guarantees were structured around a series of metrics in

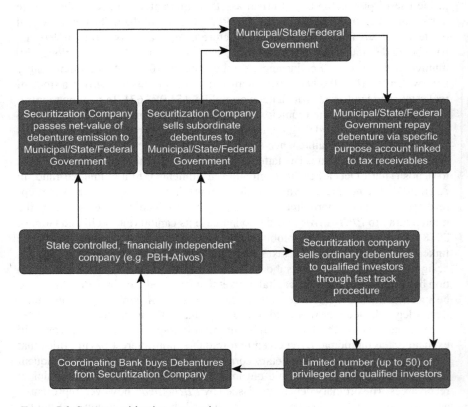

Figure 7.2 State securitization companies

Source: Author's elaboration. Based on Fattorelli and Gomes (2018).

order to control and monitor investor's risk and, if required, pass these on to the public sector. In that sense, the separate account that was created for the municipal tax receivables in guarantee of the private debentures was emblematic. The contract signed with bondholders required a liquidity ratio of 200%; that is, in any month during the implementation of the contract, it guaranteed that the outstanding balance of tax credits would at least be twice the present value amount of the non-amortized debentures in circulation (including interest and administrative charges). Moreover, in case of successful municipal negotiations of up-front payment of taxes in arrear in exchange for discounts, the contract obliged the city to immediately complement the financial resources in order to recompose the originally contracted income stream associated with the debentures.

Third, *PBH-Ativos* and its partners involved in the financial intermediation were allowed to operate according to guidelines of restricted placement of debentures in the secondary market (i.e. without spending much effort to disseminate emissions to a wider audience). This ended up stimulating an allocation device that was at the margin of the regular procedures for control and supervision from institutions such as the central bank, the monetary council and the senate. To be specific, on several occasions, the Brazilian stock exchange, through its Securities Commission, allowed a procedure that eliminated the requirement of a preliminary registering on the stock exchange, including a waiver of the standard procedure whereby the central bank is requested to pronounce itself in relation to the compliance of the prevailing norms and resolutions regarding public borrowing. The rollout of a more flexible regulatory framework for this kind of product by the Securities Commission had started in 2009 with the design of a simplified emission regime. It projected that, after a period of 90 days, up to 50 so-called qualified investors (making minimum bids of R\$ 1,000,000) could manifest interest, while increasing the number of potential investors who could be contacted from 50 to 75 during the prospection stage of the public offer (but maintaining a limit of 50 effective buyers). As a matter of fact, this institutional environment created a oligopsony whereby a few very well-informed investors could leverage their financial founder's profit (to use Hilferding's classical expression) associated with the floating and commercialization of securities. In that respect, "the rules of the game" also reinforced a lack of transparency and the potential for leverage, considering that the pricing of the emission was undertaken on the basis of a book-building method, whereby the coordinating financial institution of the emission would explore the potential demand for titles on the basis of an iterative procedure involving this small community of interested investors.[9]

Not unsurprisingly, the financial-institutional devices used by the city of Belo Horizonte through the constitution of *PBH-Ativos* generated polemics and contestation. Housing movements protested against the heterodox alienation of land in areas that were already committed to alternatives uses according to earlier negotiations. The non-governmental organization "Citizens for Debt Auditing" contested the legality of *PBH-Ativos* and its financial-institutional devices in light of constitutional clauses and the National Law on Fiscal Responsibility. Eventually, in May 2017 a municipal parliamentary inquiry was installed with the objective of investigating the legality of the institutional and financial practices of *PBH-Ativos*. The inquiry was inconclusive and filed in November of the same year, particularly in light of the disagreements among the members of the commission itself. Nevertheless, its work generated clear indications of the financial disequilibria that had been occurring during the initial years of the company. For example, a consolidated analysis of the cash flows of *PBH-Ativos* in the period between April 2014 and June 2017 suggested a preliminary municipal deficit of almost R\$ 70 million. The latter was due to the difference between the cash inflows from the city of Belo Horizonte associated with the transfer of R\$ 531 million in tax receivables to *PBH-Ativos,* and the cash outflow from *PBH-Ativos* to the municipality of R\$ 462 million (related to the revenues of a first emission of debentures, worth R\$ 200 million, which were transferred to the municipality of Belo Horizonte, a clear indication of municipal borrowing, and the transfer of effective tax receipts of R\$ 262 million back to the municipality) (Fattorelli and Gomes, 2018: 180).[10]

Irrespective of the specific contestations and polemics that have surrounded the experience of *PBH-Ativos*, several Brazilian cities and states have now adopted similar devices in order to re-connect with capital markets in the search for "solutions" in times of austerity. For instance, in October 2015 the state of Rio de Janeiro created by decree its CFSEC S/A (*Companhia Fluminense de Securitização*; that is, the Rio de Janeiro State Company for Securitization). In November 2015, the city government of Recife (in the state of Pernambuco) constituted RECDA S/A (*Companhia Recife de Desenvolvimento e Mobilização de Ativos*; that is, Recife company for the development and mobilization of assets). On December 30, 2015, the municipal council of the city of Porto Alegre (in the state of Rio Grande do Sul) authorized the city to create its municipal securitization company Invest POA (*Empresa de Gestão de Ativos do Município de Porto Alegre S/A*; that is, Company for Asset Management of the municipality of Porto Alegre). On July 16, 2015, the state legislature of Paraná approved the creation of PRSEC (*Campanhia Paranaense de Securitização*; that is, the Paraná State Company for Securitization). On July 15, 2013, the municipality of Salvador, capital city of the state of Bahia, received council approval to set up CDEMS (*Companhia de Desenvolvimento e Mobilização de Ativos de Salvador*; that is, Company for the Development and Mobilization of Assets of Salvador). Experiences in São Paulo are somewhat older, but it adopted the same institutional and financial architecture. More specifically, on December 20, 2007, the municipality of São Paulo received approval from the council to create SPDA (*Companhia São Paulo de Desenvolvimento e Mobilização de Ativos*; that is, São Paulo Company for the Development and Mobilization of Assets). Likewise, CPSEC (*Companhia Paulista de Securitização*; that is, the São Paulo State company for Securitization) was created on the October 15, 2009, after having received authorization from the state legislature. The previously mentioned list indicates a striking similarity in terms of the names of the securitization companies that were created, as well as their objectives and legal structure. With the exception of São Paulo, the proliferation of experimentation takes place after 2013, which represents a year that can be considered something of a turning point in terms of the reduced economic growth and gradually increasing political instability of the country.

In all of these experiences, municipal and state prosecutors as well as the federal court of auditors have generated objections and contestations in relation to the securitization of tax receivables and non-tax income as a mechanism to access capital markets. With the purpose to solve the legal uncertainties that have surrounded this new subnational, somewhat subversive praxis, senate proposal Nº 204, from 2016, now aims to "legalize" the securitization of tax and non-tax receivables by federal and state governments and cities, without classifying the device as (indirect) borrowing, which would clearly contradict the earlier-discussed National Law on Fiscal Responsibility. In a way, this move consolidates a contested political project aimed at framing crisis and austerity, and its solution, in terms of the requirement to release regulatory restrictions on borrowing and securitization in order to unlock the unexplored potential of capital markets.

7.3.3 Critical discussion

If, as argued by Hinkley (2017: 2125), crisis and the proposed mechanism to solve it become potentially contagious, this explains much of the relatively rapid proliferation of state-owned securitization companies which have been set up by local and regional authorities as a financial-institutional device to fix austerity. While it is outside the scope of this chapter to analyze the intriguing question of how, in each of these different places, the circulation of technical experts, consultants and their financial-legal devices has generated a relatively standardized solution, the key point here is that finance and capital markets have, in a relatively short period of time, evolved from the central problem to a strategic solution in the planning, management and governance of austerity in times of political and economic crisis.[11]

To understand this paradox, we need to revisit the framing of crisis and austerity during the 1990s and analyze how the narratives subtly transformed into something different in recent times. In the 1990s, crisis, and its opposite (i.e. macroeconomic stabilization and inflation control) were framed around the need for a rollback of the state and developmental finance. Fiscal discipline, austerity and the extinction of state banks represented pre-conditions to reduce the crowding out of private savings and investments and the gradual "buildup of solid capital markets". In a way, the National Law on Fiscal Responsibility represented the institutional crown jewel that consolidated the politics and "the ownership of the crisis" (Bayırbağ et al., 2017) that underpinned the particular shape of the neoliberal project during the 1990s.

In the last decade or so, particularly from 2013 onward, crisis and austerity have re-merged, albeit in a somewhat different framing.[12] In the contemporary setting, finance and capital markets are seen as disconnected from the broader demands for funding in light of cumbersome regulatory restrictions and lack of guarantees that tend to increase systemic investor risk. Enabling state rollout and regulatory streamlining are now constructed as essential building blocks in providing austerity solutions through the connection with "existing capital markets", which allows to tap into its financial resources and know-how in the efficient planning and management of assets. At the same time, the disputes around the legal status of the municipal and state securitization companies, as well as the uncertainty regarding the pending approval of senate proposal Nº 204/2016, indicate that this particular "politics of crisis", framed around the lack of entanglement between cities and efficiently working capital markets and finance, is still under construction.

Brazilian state rescaling has reflected a specific "politics of austerity". The 1990s were characterized by a framing of crisis that privileged the downloading of austerity to cities and states, as well as a centrally imposed cut-off from their developmental, politically mediated access to finance, which would block the emergence of consolidated private capital markets. More recently, however, cities and states have claimed their access to subnational borrowing as a solution to austerity, within an institutional environment they nevertheless perceive to be as excessively restricted in terms of tapping into credit markets. The rapid proliferation of a "subversive" subnational praxis of connecting with capital markets has now put pressure to restructure and reframe the existing national legal framework that guides issues such as fiscal responsibility, access to borrowing and securitization.

Finally, it remains to be seen whether the so-called solutions that have been designed and adopted through municipal and state-controlled securitization companies represent effective value added in terms of bringing in the financial resources and professional know-how that was promised when signing the contracts. In the case of *PBH-Ativos*, for example, the parliamentary inquiry generated substantial doubts regarding the financial advantages for municipal management. Moreover, the transfer of municipal assets such as tax receivables into a separate specific-purpose fund merely served as a device to provide guarantees and reduce risk for bond-holders, while city government effectively maintained responsibility for the planning and management of delinquent taxes or taxes in arrears.

7.4 Conclusion

The analysis in this chapter allows for three conclusions, each of them providing perspectives for further research on urban austerity and crisis. First, the investigation of the trajectory of state spaces in the making and breaking of finance and capital markets provides important insights to a better understanding of the entanglements between crisis and austerity itself. In the 1990s, the state and developmental finance required to be rolled back in a clear framework of crisis and austerity whereby finance, being constituted as essentially "public" through the prevailing state banks and subsidized compulsory savings, was crowding out badly needed private investment in order to regain macroeconomic momentum and consolidate inflation control by the Plano Real. In the post-2013 framing of crisis and austerity, however, enabling regulatory rollout is perceived as instrumental in connecting with capital markets and finance, which are being represented as efficiently working "private" institutions, in order to augment public investments in cash-strapped cities and states. To some extent, this explains the rapid dissemination of experiences such as *PBH-Ativos*, considering its grounding in a logics and politics of austerity whereby capital markets are represented as already being out there, and require regulatory streamlining in order to unlock their potential in terms of the available savings and technical know-how that are instrumental to regain public investment and growth. The Brazilian case also suggests the potential for further research on the role of state spaces in the making/breaking of the nexus between crisis-austerity and thin capital markets in the geographical and historical trajectory of different countries and cities, particularly in the setting of the Global Urban South.

Second, the Brazilian experience clearly indicates that scaling and rescaling processes are part of what might be called the "politics of austerity" (Bayırbağ et al., 2017: 2026) in terms of the disputes that accompany the meaning-making and owning of any crisis. Federal downloading of a specific solution (i.e. subnational fiscal austerity without developmental finance) represented the dominant strategy of the 1990s. The post-2013 crisis has now generated a series of state-spatial tensions aimed at up-scaling quite a different solution (i.e. connecting with finance and capital markets). While it was outside the scope of this chapter, these findings also suggest the potential to link the urban crisis-framing literature, which was summarized in the introductory section, and the debates that have been undertaken on the issue of policy mobility (McCann and Ward, 2013), and "the production of cities through global-relational connections" (Baker and Temenos, 2015: 825). More specifically, the idea of policy mobility has potential in exploring how specific devices such as municipal and state securitization companies "travel" and become contagious to the point of influencing the national debate regarding the need for regulatory reform.

Finally, our preliminary reading of the rescaling and restructuring of the capital market-austerity nexus in the specific Brazilian setting suggests that much more detailed work is required in order to flesh out the complex interfaces between the governance of securitization and the securitization of governance. More specifically, the institutional and financial devices used in the making of urban securitization in Brazilian cities and states go beyond the common domain of real estate and require delving into the variegated projects and strategies that are used to transform the public fund itself (i.e. tax and non-tax receivables alike, as well infrastructure services) into a bundle of contracted income streams that can be used to access presumably efficient capital markets and solve austerity. However, the initial evidence of the ongoing experiences of municipal and state securitization indicates not only legal controversies but also significant financial contradictions. As such, there is an urgent need to flesh out the financial constitution and legitimization of these operations as a curiously "performative" device in the making of risk-averse and weak capital markets, as well as their implications for the gradual hollowing out – from the inside by the state itself as the *ersatz* of finance capital (Oliveira, 1998) – of shared democratic governance, and its filling in by post-political, bond-holder arrangements for the planning and management of cities.

Notes

1 In addition, in Chapter 1 we discussed specific Post-Keynesian literature on the relations between risk, urban governance and capital markets, according to which city governments would gradually move from stable hedge finance, to increasingly speculative Ponzi-type financial strategies in municipal markets for bonds and debentures. See Kirkpatrick (2016).

2 Approved by the National Monetary Council ruling N° 162 in 1995.

3 More specifically, federal debt increased from R$ 60.7 billion in July 1994 to R$ 624.1 billion in December 2001. Forty-seven per cent of this increase, that is, R$ 297.7 billion, was due to re-financing municipal and state debts (Nascimento and Debus, 2001: 46).

4 The figures are taken from Costa (2012: 181–182). The values are converted according to the overvalued currency that was prevailing during the second half of the 1990s. This author also recalls that the extremely high interest rates prevailing after the Plano Real led to escalating costs associated with the delay in negotiations. For instance, if a re-finance agreement would have been signed two years earlier, the debt involved in the re-negotiation would be approximately half the amount.

5 As will be discussed subsequently, this clause would generate one of the loopholes utilized by companies such as *PBH-Ativos*.
6 This empirical section is based on a detailed analysis of the experience of the *PBH-Ativos S/A* company in the city of Belo Horizonte (state of Minas Gerais), using archival research, an investigation of the debenture contracts and the available literature on the case, particularly by authors such as Canettieri (2017) and Klink (2018). The contracts are open-access public data and available at: http://pbhativos.com.br/transparencia/contratos-e-convenios/ (accessed: June 1–30, 2018).
7 Source: open source site from the company *PBH-Ativos* available at: http://pbhativos.com.br/a-pbh-ativos/quem-somos/ (accessed July 26, 2019).
8 According to article 2 from Law N° 10.003, November 25, 2010, which created the company *PBH Ativos S/A*.
9 More specifically, by using an initial price range, the coordinating bank explores whether the demand, as revealed by potential investors, covers the financing requirement of the public sector. If investor demand is lower than the requirements for finance, the price that was initially proposed proved too high and needs to go through successive "trial and error" adjustments in order to match funding needs and investors demand.
10 This deficit of R$ 70 million reflects financial gains of R$ 28.3 million of the coordinating bank *BTG Pactual Ltd*, while the rest is internalized by *PBH-Ativos* itself as part of its overall cash flow management.
11 The specific experience of *PBH-Ativos* also indicates the importance of "transfer agents" (Baker and Temenos, 2015; McCann, 2011) in the circulation of experiences. In that sense, it is worthwhile to mention the role of Edson Ronaldo Nascimento (ex-president of *PBH-Ativos*), who worked as a finance secretary in the states of Goías and Tocantins and in the federal district of Brasilia. He was also consultant to the IMF and senior staff of the Brazilian Ministry of Finance (CNACD, n.d.; Canettieri, 2017).
12 For instance, see the national newspaper coverage from *Folha de São Paulo*, on July 28, 2019, which was titled: "In an attempt to solve fiscal crisis, seventeen states prepare privatizations". Available at: www.folha.uol.com.br/mercado/2019/07/na-tentativa-de-superar-crise-fiscal-17-estados-preparam-privatizacoes.shtml (accessed July 28th, 2019).

References

Addison F and Artioli F (2019) Four type of urban austerity: Public land privatisations in French and Italian cities. *Urban Studies*. Online First. Available at: https://doi.org/10.1177/0042098019827517 (Accessed June 17, 2019).

Affonso BA de and Silva PLB (eds.) (1995) *A Federação em perspectiva: Ensaios selecionados*. São Paulo: FUNDAP.

Alke J (2018) Authoritarian neoliberal rescaling in Latin America: Urban in/security and austerity in Oaxaca. *Globalizations*: 1–16.

Arampatzi A (2017) The spatiality of counter-austerity politics in Athens, Greece: Emergent "urban solidarity spaces". *Urban Studies* 54(9): 2155–2171.

Armondi S (2017) State rescaling and new metropolitan space in the age of austerity: Evidence from Italy. *Geoforum* 81: 174–179.

Baker T and Temenos C (2015) Urban policy mobilities research: Introduction to a debate. *International Journal of Urban and Regional Research* 39(4): 824–827.

Bayırbağ MK; Davies JS and Münch S (2017) Interrogating urban crisis: Cities in the governance and contestation of austerity. *Urban Studies* 54(9): 2023–2038.

Brenner N (2004) *New state spaces: Urban governance and the rescaling of statehood*. Oxford: Oxford University Press.

Brenner N; Peck J and Theodore N (2010) Variegated neoliberalization: Geographies, modalities, pathways. *Global Networks* 10(2): 182–222.

Canettieri T (2017) A produção capitalista do espaço e a gestão empresarial da política urbana: o caso da PBH Ativos S/A. *Revista brasileira de estudos urbanos e regionais* 19(3): 513–529.

CNACD (Coordenação Nacional da Auditoria Cidadã da Dívida) (n.d) *Projetos cifrados. PLS 204/2016, PLP 181/2015 e PL 3337/2015 visam "legalizar" esquema fraudulento.* Brasília: CNACD.

Costa FN da (2012) *Brasil dos bancos*. São Paulo: EDUSP.

Engelen E and Glasmacher A (2018) The waiting game: How securitization became the solution for the growth problem of the Eurozone. *Competition & Change* 22(2): 165–183.

Fattorelli ML and Gomes JM (2018) Securitização de créditos: Desvio de arrecadação e geração de dívida pública illegal. *Revista Direitos, Trabalho e Política Social* 4(7): 165–199.

Fiori JL (1999) *Estados e Moedas no Desenvolvimento das nações*. Petrópolis: Vozes.

Fuller C and West K (2017) The possibilities and limits of political contestation in times of "urban austerity". *Urban Studies* 54(9): 2087–2106.

Hall S and Jonas AEG (2014) Urban fiscal austerity, infrastructure provision and the struggle for regional transit in "Motor City". *Cambridge Journal of Regions, Economy and Society* 7: 189–206.

Hinkley S (2017) Structurally adjusting: Narratives of fiscal crisis in four US cities. *Urban Studies* 54(9): 2123–2138.

Keating M (2013) *Rescaling the European state: The making of territory and the rise of the meso*. Oxford: Oxford University Press.

Kirkpatrick LO (2016) The new urban fiscal crisis: Finance, democracy, and municipal debt. *Politics & Society* 44(1): 45–80.

Klink J (2001) *A Cidade-região. Regionalismo e reestruturação no Grande ABC Paulista*. Rio de Janeiro: Editora DPA.

Klink J (2018) Metrópole, moeda e mercados. A agenda urbana em tempos de reemergência das finanças globais. *Cadernos Metrópole* 20(43): 717–742.

Marcuse P (1981) The targeted crisis: On the ideology of the urban fiscal crisis and its uses. *International Journal of Urban and Regional Research* 5(3): 330–354.

Martí-Costa M and Tomàs M (2017) Urban governance in Spain: From democratic transition to austerity policies. *Urban Studies* 54(9): 2107–2122.

McCann E (2011) Urban policy mobilities and global circuits of knowledge: Toward a research agenda. *Annals of the Association of American Geographers* 101(1): 107–130.

McCann E and Ward K (2013) A multi-disciplinary approach to policy transfer research: Geographies, assemblages, mobilities and mutations. *Policy Studies* 34(1): 2–18.

Nascimento ER and Debus I (2001) *Lei Complementar No 101/2000. Entendendo A Lei de Responsabilidade Fiscal*. Brasília: Secretaria do Tesouro Nacional, Ministério da Fazenda.

O'Connor J (1973) *The fiscal crisis of the state*. New York: St Martin's Press.

Oliveira F de (1998) *Os Direitos do antivalor*. Petrópolis: Vozes.

Oosterlynck S and Gonzales S (2013) "Don't waste a crisis": Opening up the city yet again for neoliberal experimentation. *International Journal for Urban and Regional Research* 37(3): 1075–1082.

Peck J (2017a) Transatlantic city, part 1: Conjunctural urbanism. *Urban Studies* 54(1): 4–30.

Peck J (2017b) Transatlantic city, part 2: Late entrepreneurialism. *Urban Studies* 54(2): 327–363.

Peck J and Whiteside H (2016) Financializing Detroit. *Economic Geography* 92(3): 235–268.

Tavares MdC and Fiori JL (eds.) (1997) *Poder e Dinheiro. Uma Economia política da globalização*. Petrópolis: Vozes.

Conclusion

This book has provided three contributions to the existing literature on financialization. First, instead of analyzing financialization *in* cities, we have argued that research should move into the direction of investigating financialization *of* cities and *through* cities (as urbanization of finance). This implied looking at urban space as more than a receptacle of global finance. Our approach was to investigate the entanglements between urban space and finance from the perspective of cities' role as privileged, contested arenas associated with the generation, circulation and consumption of value as well as the reproduction of daily life. Cities represent a prime analytical and empirical lens in order to articulate an understanding of finance in terms of what it is, and how it is socially constituted, on the one hand, and what it tends to do, on the other.

On the basis of this perspective, we have worked out variations on a theme according to which cities are actively involved in the shaping of variegated patterns of financialization, either through the monetization of additional building rights in large redevelopment projects (Chapter 4), by filling in particular calculative practices in the pricing and valuation of water companies and shaping relations of debit-credit in basic sanitation (Chapter 5), by capitalizing subsidies from national housing programs into prices for land and housing (and, as such, hollowing out these program themselves) (Chapter 6) and, last but not least, by reframing the austerity-capital market nexus in times of crisis (Chapter 7).

While looking at the urban question as a scalar question (Brenner, 2000, 2004), in all of these examples we have revisited critical urban studies and the contemporary literature on the contradictory role of the state in the (re)production of space. Nevertheless, we have provided new analytical insights by articulating the existing state spatial and scalar literature (which – with few exceptions – is rather silent on finance) with heterodox thinking on the non-neutral and potentially de-stabilizing role of money as a social relation (which tends to marginalize space).

Under the prevailing international monetary governance, marked by central banking systems and (supra)nationally scaled regulation, cities have evidently since long lost their powers to directly emit and influence the circulation of money. Nevertheless, they can be creative and flexible producers of all sorts of unregulated credit and fiduciary money, as was discussed in our example of securitized building rights and building quotas, which are used by cities to pay developers and contractors (Chapter 4). Likewise, when analyzing urban securitization (Chapter 7), we saw that cities are actively involved in filling in and, if required, stretching their relationship with capital markets beyond what is projected in prevailing regulatory frameworks.

Moreover, scaling and rescaling are key for a better understanding of the politics of the making (or breaking) of financialization and access to money, credit and finance. This link was clear in our review of the federal downloading of austerity and borrowing restrictions in the 1990s and the recent pressure to reframe and upscale access to capital markets as a solution to crisis and austerity (Chapter 7). In the other experiences we discussed, the local-metropolitan experimentation was also embedded in the broader, contested state spatial politics of scale regarding the transformation of city space into contracted streams of income.

In relation to basic sanitation, for instance (Chapter 5), the conflicts between state mixed-capital listed company SABESP and several of the cities regarding the level of debt of municipal utilities represented a dispute over how to fill in the metropolitan project and strategy for the sector in general, and the control over wholesale and retail markets in particular, with SABESP mainly interested in providing a stable and predictable stream of tariff-income for its shareholders (both the state itself and private stockholders). At the time of writing this chapter, this institutional-financial battlefield for metropolitan São Paulo is being up-scaled through a federal proposal to review the national framework legislation with the clear objective to stimulate outright privatization or a more aggressive entrance of the private sector in agreements with municipal and state companies regarding concessions and public-private partnerships.

Large urban redevelopment projects also reflect the scalar politics of money and finance in the transformation of space. The case of inner-city harbour redevelopment in Rio (more particularly, the *Porto Maravilha* project, which was briefly discussed in Chapter 4) was emblematic. The audacious, albeit complex financial-institutional engineering, which involved securitized building certificates (CEPACS) that were linked to real estate investment funds and a public-private partnership to build, operate and run public transportation, could not have taken off (neither could it have landed rather hard for that matter) without the regulatory easing regarding the use of the resources from the national social housing fund *FGTS* (sourced by the compulsory contributions out of wages) in the up-front multi-billion acquisition of CEPACs for the redevelopment project.

Likewise, in Chapter 6 we discussed the critical evaluation studies regarding the project and market-driven approach of the national social housing program MHML. The program was considered to hollow out the political project of urban reform, which was framed around cities' efforts to design and implement redistributive land-use planning aimed at the provision of affordable, well-located housing for the poorest of the poor (Target Group 1). In the specific case of the Greater ABC region that was examined, however, the previous demarcation of Special Social Interest Zones in already-serviced locations effectively enabled cities to quickly benefit from the Target Group 1 resources that were provided by

MHML; at the same time, however, the lack of cities' effective leverage over booming land markets provided incentives for developers to capitalize subsidies and build outside the price limits and subsidy guidelines as established for the "not so poor" segments of the program (i.e., Target Groups 2 and 3). In a way, then, if the national MHML program was hollowing out local urban reform, the opposite was also true; pricing and product decisions by developers generated "money running uphill" and a gradual "portfolio shift" of the housing units that had initially been contracted within the limits of the program to middle-income groups, which acquired them through market-based finance.

Finally, financialization of cities, and through cities, has implications for research on urban and metropolitan governance. Here we have only started to touch upon this theme. More specifically, the momentum and "virtuous cycle" (Maricato, 2011; Klink and Denaldi, 2015) that accompanied Brazilian planning theory and progressive praxis after re-democratization until recently consolidated a narrative structured around the intrinsic value of collaborative and communicative approaches and shared territorial governance (Healey, 2003; Fainstein, 2000; Fainstein and Campbell, 2012). Our work suggests that this communicative-collaborative turn has proven fragile in light of a historical trajectory of large social-spatial disparities and techno-bureaucratic, authoritarian and clientelist planning and management. Instead, what seems to be emerging in the cases we have examined is a pattern of urban shareholder governance, which provides a potentially fertile ground to consolidate post-political, neoliberal planning and management (Purcell, 2009).

This brings us to the book's second contribution. There is nothing particularly inherent about the transformation of cities, as use values and privileged spaces for the reproduction of daily life, into a portfolio of tradable, income-yielding assets. We have argued that an analytical articulation of political economy (with its emphasis on investigating the contradictory effects of finance in terms of exploiting us without producing) with social studies of finance and work on performativity (with a conceptual focus on what finance is and how it is constituted) provides useful insights into this complex transformation of city spaces. More particularly, we have explored the potential of a political economy of valuation in understanding the contested and variegated making of urban financialization, as well as the complementarities between what finance is and does in cities.[1]

Planning praxis and theory is crucial for this kind of mediation to succeed. As a matter of fact, our concept of "performative urbanism" indicates that planning becomes decontextualized (Savini and Aalbers, 2016) in the sense that it is actively involved in the utilization of models of corporate finance with the purpose of pricing and valuating the city as an asset. Moreover, planning is mobilized in order to put these models to work according to the political projects and strategies of specific actors. Performative urbanism represents something of a mid-level concept we have used in order to articulate the empirical work of this book and some of the broader theoretical claims that have emerged on urban financialization and the urbanization of finance.

As a consequence, it becomes important to flesh out the relations between planning and modelling as well as the framing, meaning-making and the practices involved in the articulation between cities and finance. Along the book, we have presented some of the "technicalities" as well as discursive dimensions that accompany these processes. In relation to the former, for example, we have discussed how planners mobilize technical tools and devices such as assetization, (differential) capitalization and how the valuation of risk is entangled with the capture of public funds. This sheds light on the particular circulation of norms, conventions and calculative practices that contributes to the making of financialized cities.

But models require social actors to put them to work in order to transform city space; that's where the discursive-ideational dimensions unfold. In a way, planners and their partners in professional networks and communities contribute to "un-cage" the models to do their work. This was true in the utilization of securitized building rights in large redevelopment projects, which, according to some of the brokers we interviewed, contributed to a newly emerging Brazil, with awareness and responsibilities regarding the positive role of capital markets. As a matter of fact, the PIUs (*Projeto de Intervenção Urbana*, meaning Project for Urban Intervention) and urban redevelopment projects were discursively framed as ideal devices to articulate broader master plans and zoning laws, traditionally perceived as having generated little practical results in the Brazilian planning trajectory, with a badly needed effective transformative praxis that would be able to mobilize the (financial) resources and creativity of the communities involved. In a way, then, the narratives behind urban projects in general, and project finance in particular, were key in minimizing the risk of private partners and earmarking projected revenues within the perimeters of specific interventions. The water pricing narratives we discussed were likewise powerful. To be clear, companies such as SABESP adopted a financial argument, according to which pricing represented the superior mechanism to provide the sustainable and universal right to water in line with the UN objectives; at the same time, however, it proved to be a key metric that was able to generate (intergovernmental) debts, which should be paid by municipal utilities in one way or another. Along the same lines, state securitization companies contributed to the proliferation of a narrative on finance as a solution in times of crisis and austerity, which would unlock the potential of capital markets in terms of the efficient management of the city's assets and tap into new financial resources and professional know-how.

The approach that was adopted in this book in terms of a closer articulation between political economics and social studies of finance around a political economy of urban valuation suggests at least two elements for a broader research agenda along these lines.

First, there is need for a better understanding on how specific devices, norms and conventions, as well as pricing and valuation techniques "travel" and contribute to the surprisingly rapid dissemination of contradictory practices that accompany the variegated making of urban financialization. We have seen several illustrations of this circulation, ranging from the state-controlled securitization companies in São Paulo and Belo Horizonte, which spread to several states and capital cities; the financial-institutional engineering behind the Rio de Janeiro inner harbour redevelopment project that has inspired the design of similar operations in other cities; as well as SABESP's approach to water pricing that has disseminated similar arrangements for shareholder governance, which are now behind the pressure aimed at national regulatory reform.

In order to come to grips with this apparently contagious nature of the dissemination of models, practices and devices, it is required to link up with the geographically inspired work on policy mobility (Baker and Temenos, 2015; McCann, 2011; Peck and Theodore, 2015; Robinson, 2015), with its emphasis on the relational, multi-scalar making of urban space (see Chapter 7), as well as with the literature on the "travelling" of planning ideas, as developed by authors such as Patsey Healey (2011, 2013). The articulation with these literatures will also contribute to re-constitute certain "origin narratives" behind specific planning practices and financial devices, as well as to explain how, why and what changes once they "travel", get disseminated and "planted" somewhere else.

This brings us to a related, second point for further research on the political economy of urban valuation and pricing. Much of the policy mobility and travelling of planning ideas and financial-institutional devices effectively represent a transnational flow and circulation of norms, conventions and practices that occur in a context that does not fit (anymore) the containers of the central and peripheral national space economies. Thus, more international work on cities is required, which is nevertheless grounded in an analytical and empirical framework that takes the asymmetric interdependencies of the prevailing global monetary order seriously, while recognizing that the making and dissemination of financialized calculative practices in cities of the Global South are likewise a key element in understanding the legitimization and maintenance of this international arrangement in the first place.

This leads us to the book's third contribution to the research agenda on financialization. While recognizing that central banks, monetary authorities and macroeconomic current and capital account and exchange relations evidently continue to be important objects of academic research, this book has argued in favour of conceptualizations that go beyond the usual macro-structural metaphor of financial centre and periphery, which was adapted from the classical dependency theory. To be specific, there are important state spatial and scalar tensions within national space economies that unfold in cities and metropolitan arenas, within an increasingly interdependent monetary order. This requires approaches that stretch our imaginaries beyond the non-spatial containers of macroeconomic accounting statistics. Investigating cities in the Global South sheds light on the constitution and reproduction of the global asymmetric pattern of financialization itself.

While more detailed work along these lines is required, on the basis of the Brazilian case we have generated an initial hypothesis that an analytical reading of the nexus between cities and capital markets that is sensitive to the historical and geography trajectories of specific countries provides insights into understanding the contemporary re-emergence of global finance.

The Brazilian early urbanization between the second half of the nineteenth century and the beginning of the twentieth century was not accompanied by efforts to design and strengthen local capital markets in order to facilitate the allocation of domestic loanable funds to finance the countries' initial stages of industrialization. In an historical context whereby the abolishment of slavery triggered a "portfolio-switch" of assets into land within the prevailing export-led coffee economy, the emerging cities were gradually transformed into privileged spaces for the extraction of urban founder's profits in real estate and urban infrastructure by national and foreign investors alike. This was something like an early urban financialization, marked by the proliferation of speculative rental housing and floating of existing companies on the foreign stock exchange. Interestingly, at that time the country was not lacking savings or loanable funds as such, but industrial investment opportunities that could have both beaten the short-term returns obtained from early urban-biased financial founder's profits, as well as unlocked an alternative development trajectory. Nevertheless, as discussed in Chapter 2, industrialization did take place, in spite of the lack of consolidated domestic markets for long-term credit and shares.

The rise and fall, during 1930–1985, of the developmental state, which concentrated its efforts around the planning and finance of industrialization, urbanization and the building of a national space economy, didn't significantly change this fact; despite some (failed) efforts to create mortgage institutions and the significant presence of state-owned banks, the country nevertheless did not consolidate deep, liquid and long-term private markets for credit and shares. By and large, industrialization was financed through a mix of state funds, foreign investments and reinvested profits, which were particularly relevant in family-owned building and construction companies and contractors. In the meantime, the role of the public fund in providing indirect (and insufficient) contributions to social wages, as well as to the finance of urbanization and industrialization, triggered exponential debt levels, which generated additional opportunities for the generation of financial profits.

Thus, the restructuring, rescaling and hollowing out of the developmental state, which occurred from the mid-1980s onward, put both cities, which during decades had accumulated huge socio-spatial and environmental deficits, and the public fund itself again on the radar of domestic and international investors alike.

As such, the first generation of rollback policies in the 1990s affected developmental governance and finance through the outright privatization of state banks and assets and the re-entrance of private national and international finance, which was also stimulated by the deregulation of relatively sheltered domestic banking (Chapter 3). The post-2015 economic crisis and corruption scandals have deepened the internationalization process through complementary mechanisms. As discussed in Chapter 4, the accusations of large-scale corruption in the context of the operation Carwash *(Lava Jato)* have dramatically affected large, family-owned and predominantly national infrastructure companies and contractors, which have become the object of acquisition efforts by international firms since 2016.

Therefore, the more recent rollback of the state and "come-back" of international finance is most likely again to prioritize cities, this time with real estate, infrastructure networks and public funds as collateral in credit operations, which are reinforced by the contemporary multi-scalar "regulatory easing" aimed at unlocking the presumable potential of efficient capital markets.

As such, unlike traditional metaphors of centre-periphery structured around national space economies, urban and metropolitan areas themselves will increasingly represent arenas that are directly contested by both national and international investors, as well as by alternative projects that denounce the drainage of the public fund and the extraction of financial profits from the city by "really existing" thin and risk-averse capital markets, and demand social justice and a fairer reproduction of daily life in the contemporary metropolis. At the time of writing, however, the prospects for such differential spaces and vitality for renewed rounds of "insurgent citizenship" (Holston, 2009) seem quite remote, at least in Brazil, particularly when looking at the impressive velocity of the meltdown of both redistributive-reformist and collaborative-participatory institutions that have been built in and by cities since the re-democratization and decentralization of the mid-1980s. As a matter of fact, it seems that the framing and meaning-making of insurgency itself are increasingly becoming the objects of political struggles (Miraftab, 2009; Randolph, 2008). In that sense, in relation to the Brazilian setting, the latest round of crisis and austerity politics have clearly announced finance-capital's multi-scalar, post-political project of framing insurgency as a strategy against the presumably corrupt, inefficient and non-transparent governance and funding of the developmental state, which represents a discourse likely to find fertile ground in the Brazilian multi-layered institutional legacy of colonial, authoritarian and technocratic planning and management of urban space.

Note

1 As discussed in Chapter 2, a political economy of valuation is concerned with the social power structures and conflicts involved in the circulation of norms and conventions that are used in the pricing and valuing of assets. It avoids falling into the methodological trap of some of the work of performativity in terms of doing "the dirty work" and naturalizing neo-classical economics. See also: Christophers (2014).

References

Baker T and Temenos C (2015) Urban policy mobilities research: Introduction to a debate. *International Journal of Urban and Regional Research* 39(4): 824–827.

Brenner N (2000) The urban question as a scale question: Reflections on Henri Lefebvre, urban theory and the politics of scale. *International Journal of Urban and Regional Research* 24(2): 361–378.

Brenner N (2004) *New state spaces: Urban governance and the rescaling of statehood.* Oxford: Oxford University Press.

Christophers B (2014) From Marx to market and back again: Performing the economy. *Antipode* 57(1): 12–20.

Fainstein SS (2000) New directions in planning theory. *Urban Affairs Review* 35(4): 451–478.

Fainstein SS and Campbell S (eds.) (2012) *Readings in planning theory.* West Sussex: Blackwell Publishers.

Healey P (2003) Collaborative planning in perspective. *Planning Theory* 2(2): 101–123.

Healey P (2011) The universal and the contingent: Some reflections on the transnational flow of planning ideas and practices. *Planning Theory* 11(2): 188–207.

Healey P (2013) Circuits of knowledge and techniques: The transnational flow of planning ideas and practices. *International Journal of Urban and Regional Research* 37(5): 1510–1526.

Holston J (2009) Insurgent citizenship in an era of global urban peripheries. *City & Society* 21(2): 245–267.

Klink J and Denaldi R (2015) On urban reform, rights and planning challenges in the Brazilian metropolis. *Planning Theory* 15(4): 402–417.

Maricato E (2011) *O impasse da política urbana.* Petrópolis: Vozes.

McCann E (2011) Urban policy mobilities and global circuits of knowledge: Toward a research agenda. *Annals of the Association of American Geographers* 101(1): 107–130.

Miraftab F (2009) Insurgent planning: Situating radical planning in the Global South. *Planning Theory* 8(1): 32–50.

Peck J and Theodore M (2015) *Fast policy: Experimental statecraft at the thresholds of neoliberalization.* Minneapolis: University of Minnesota Press.

Purcell M (2009) Resisting neoliberalization: Communicative planning or counter-hegemonic movements? *Planning Theory* 8(2): 140–165.

Randolph R (2008) A nova perspectiva do planejamento subversivo e suas (possíveis) implicações para a formação do planejador urbano e regional—o caso brasileiro. *Scripta Nova* 12(270): 98. Available at: www.ub.edu/geocrit/sn/sn-270/sn-270-98.htm

Robinson J (2015) "Arriving at" urban policies: The topological spaces of urban policy mobility. *International Journal of Urban and Regional Research* 39(4): 831–834.

Savini F and Aalbers M (2016) The de-contextualisation of land use planning through financialisation: Urban redevelopment in Milan. *European Urban and Regional Studies* 23(4): 878–894.

Index

Aalbers MB 1–2, 5, 8, 32
Administrative Council for the Protection of the Economic Order *see* CADE
agencements 30–31
Aglietta M 6, 8, 16–17, 23–24, 41, 54–56, 95
Allen J 3, 35, 59, 145
Alonso W 18
anti-value 57–58, 62–63, 82, 104n27
Arrighi G 21
ARSESP 135–137, 140–141, 143
"as efficient competitor" 137–**138**
assetization 59–61, 141–**142**, 211
asymmetric interdependency 74
Attuyer K 33, 107
austerity: crisis-austerity nexus 187; narratives 187–188, 193; urbanism 38, 188, 191, 193, 195
average incremental cost 140

Bahl RW 40
Bakker K 3–4, 35, 37
Baran PA 21
Baumol WJ 30, 60
Bayliss K 4, 35
Berle AA 26–27
Bichler S 20, 61, 78n14
Bielschowsky R 41, 66, 92
Black F 17, 30, 61
Bolaffi G 93, 104n16
"bondholder value" 38
Bonduki N 92
book-building method 201
BOVESPA 110, *136*
Brazilian savings and loans system *see* SBPE
Brenner N 3, 5, 34, 51–52, 192, 208
building right: additional 107, 109, 126, 128n17, 208; certificates 34; quota 126

building right certificates *see* CEPAC
"buyer of last resort" 73

CADE 132, 137, **139**–141, 143, 145n5, 146n14
Caixa Econômica Federal 95, 161, 173
calculative practices *see* political economy of valuation
Callon M 2, 29–30, 33, 132
Canettieri T 4, 38, 198, 206n6, 206n11
Cano W 86
capital asset pricing model 17, 56, 61
capital switching 21, 29, 42n4
Castells M 5, 51
CEF *see* Caixa Econômica Federal
centre-periphery 6, 73, 214; dependency model 41; metaphor 41, 66; model 41, 66
CEPAC 9, 107–**112**, 114–122, 125–126, 209
CEPAL 66
"cherry-picking" 35
Chesnais F 1, 21
Chiapello E 2, 30, 141
Christophers B 2–3, 19, 30–33, 39, 56–57, 132, 214n1
commensurability 40, 52, 61
common property resources 60
community infrastructure *174–177*, 181
compulsory savings fund 96–98, 151, 180
Comunidades Autônomas 192–193
concessão urbanística see urban concession
'confined economists' 30
contagion 17, 19, 26, 56
contracted income streams 37, 58, 64, 205
Costa FN da 77n6, **85**, 87–91, 102, 103n3, 195, 205n4
coupon pool capitalism 26, 28
credit multiplier 25

critical management studies 1, 19, 26
cultural economics 29

debt-deflation 26
dependency theory 6, 41, 66, 212
developer contributions *see* CEPAC
differential capitalization 20, 62, 78n14
differential rent 60, 89, 126
DiPasquale D 18, 108, 150
discounted cash flow 60, 64–66, 111, 126, 140, **142**–143
discourse theory 190–191
disintermediation 21, 28
dollar standard 54
Dymski GA 52, 73, 77n5, 103

'economists in the wild' 30, 33
Environmental Sanitation Services of Santo André *see* SEMASA
Erturk I 26–27, 43n12–13

fantasmatic logics 188, 191
Faulhaber GR 30, 60
FGTS *see* compulsory savings fund
fictitious capital 62, 64, 77n13
fiduciary alienation 96, 98, 159
Fields D 127
financial leverage 21, 42n3, 63, 78n15
Fiori JL 76, 188
founder's profits 60, 64–65, 73, 76, **84**, 88, 98, 102, 103n7, 201, 213
Froud J 26–27, 43n12–13
Fundo de Garantia de Tempo de Serviço *see* FGTS

Gotham K 32, 40
governance: collaborative **133**–134, 194; multi-level 192; shareholder 1, 10, 22, 28, 127, 135, 144, 210, 212; territorial 34, 127, 210
Graham S 3, 34–35

Haila A 3, 32
Halbert L 33, 107
Harvey D 3, 5, 7, 18, 29, 32–33, 42n4, 51, 60
hedge finance 26, 205n1
Helleiner E 31, 54–55
Hildyard N 4, 37, 144
Hilferding R 1, 20, 43n12, 63, 89, 201
human right to water 35–36

indexation **84**, 90, 93
insurance fund 95

"insurgent citizenship" 7, 10, 214
internal rate of return 22, 25, 59, 64, 78n18, **113**–114
IRR *see* internal rate of return

Jensen M 27
Jessop B 53, 77n1
Johal S 26–27, 43n12–13

Kaika M 33, 42n3
Keating M 187, 192
Keynes JM 24–26, 54

land banks 3, 33, 98
land-value capture 110
Lapavitsas C 1, 5, 21–22, 39, 62, 66, 68, 77n13
Latour B 29–30
Lava Jato 121, 128n11–12, 214
Lazonick W 27
Leaver A 26–27, 43n12–13
Lefebvre H 5, 51, 86
Linn JF 140
Lipietz A 22
liquidity preference 25
loan to value ratio **154**, 181

Maastricht Treaty 196
MacKenzie D 1–2, 29–30, 132
Maleronka C 108–110, 119, 122, 128n4
marginal efficiency of capital 25
margin squeezing 137, 145n9
Markowitz HM 61
Marvin S 3, 34–35
Means GC 26–27
Meckling W 27
Minsky H 1, 24–25, 38, 64
monetarism 17, 25
money: commodity 53–55, 77n6, **84**; credit 25, 41, 53–55, 95, 107, 109, 126; endogenous 17, 24–25, 42n1, 43n10, 55; fictive 53; fiduciary 55, 88, 208; "high-powered" 25; neutrality of 17, 19
moral hazard 19, 32, 183n21
Muth R 18
Myrdal G 41, 66, 86

National Housing Bank *see* Caixa *Econômica Federal*
national system for sanitation *see* PLANASA
net present value 2, 60, 62–63, 77n13, 114, 140–141, 146n15, 199
new institutional economics 23, 60

Nitzan J 20, 61, 78n14
NPV *see* net present value

O'Connor J 186
Oliveira F de 57, 86, 101, 104n27, 186, 205
operation Carwash *see Lava Jato*
O'Sullivan M 27

Paulani LM 8, 16, 41, 66, 94
Peck J 6, 38, 52–53, 59, 64, 187, 193
"performative urbanism" 64, 65, 210
performativity 1–2, 5, 28, 30–31, 39,
    43n15–16, 64, 210, 214n1
PIU *see* Urban Intervention Projects
PLANASA **85**, **133**–134
planning: culture 7, 11n1;
    "decontextualization of" 125; as a
    "praxis" 77n3
Polanyi K 2, 23, 38
policy mobility 205, 212
political economy of valuation 31, 53, 59,
    64, 131, 210, 214n1
Ponzi finance 26
port project *see Porto Maravilha*
Porto Maravilha 111–**112**, 115, 119–120,
    209
post-colonial studies 52
price-cap 140–141, 143
price elasticities 18, 144
principal-agent 1
Pryke M 3, 35, 59, 145
PT *see* Worker's Party

quantity theory of money 17

rational expectations 16–17, 19
Real Estate Finance System *see* REFS
REFS 96–98
Regulatory Agency for Sanitation and Energy
    of the State of São Paulo *see* ARSESP
reverse causation 24–25
Rolnik R 32, 127
Rugierro L 33, 42n3

Sanfelici D 3, 33, 98
SBPE **85**, 161, 178
Scholes M 17, 30, 61
Securities and Exchange Comission *see*
    BOVESPA
securitization company 198–199, *200*, 202
SEMASA 10, 131–132, 137–141, 143–144,
    146n14
Smolka M 109, 116
social wage 22, 57, 100, 102, 104n27, 180,
    186, 213
solo criado see building right
special social interest zones 154, 169,
    171–172, 209
Stiglitz JE 74
Sweezy PM 21

Tawney RH 26
tax increment finance 4, 37, 107, 190
TIF *see* tax increment finance

"un-collaborative commodities" 58
United Nations Economic Comission for
    Latin America and the Caribbean *see*
    CEPAL
UPO *see* Urban Partnership Operations
urban commons 3
urban concession 124
urban entrepreneurialism 31, 33–34, 51,
    145
Urban Intervention Projects 108, 128n15
Urban Partnership Operations 9, 34, 107,
    112, 190
urban securitization 186, 205, 209

Weber R 4, 37, 107
Wheaton W 18, 108, 150
Williams K 26–27, 43n12–13
Worker's Party 7, 102, 121–122, 148, 150
world money 24, 41, 54, 67–68, 73–74

ZEIS see special social interest zones
Zwan vd N 1

Printed in the United States
by Baker & Taylor Publisher Services